Combat Search and Rescue in Desert Storm

Darrel D. Whitcomb
Colonel, USAFR, Retired

Air University Press
Maxwell Air Force Base, Alabama

September 2006

Air University Library Cataloging Data

Whitcomb, Darrel D., 1947-
    Combat search and rescue in Desert Storm / Darrel D. Whitcomb.
       p. ; cm.
    Includes bibliographical references.
    A rich heritage: the saga of Bengal 505 Alpha—The interim years—Desert Shield—Desert Storm week one—Desert Storm weeks two/three/four—Desert Storm week five—Desert Sabre week six.
    ISBN 1-58566-153-8
    1. Persian Gulf War, 1991—Search and rescue operations. 2. Search and rescue operations—United States—History. 3. United States—Armed Forces—Search and rescue operations. I. Title.

    956.704424—dc22

Disclaimer

Opinions, conclusions, and recommendations expressed or implied within are solely those of the author and do not necessarily represent the views of Air University, the United States Air Force, the Department of Defense, or any other US government agency. Cleared for public release: distribution unlimited.

© Copyright 2006 by Darrel D. Whitcomb (Learn5252@aol.com).

Air University Press
131 West Shumacher Avenue
Maxwell AFB AL 36112-6615
http://aupress.maxwell.af.mil

*This work is dedicated to the memory of
the brave crew of Bengal 15.
Without question, without hesitation, eight soldiers
went forth to rescue a downed countryman—
only three returned.
God bless those lost, as they rest in their eternal peace.*

Contents

Chapter		Page
	DISCLAIMER	ii
	DEDICATION	iii
	ABOUT THE AUTHOR	ix
	FOREWORD	xi
	INTRODUCTION	xv
	Notes	xix
1	A RICH HERITAGE: THE SAGA OF BENGAL 505 ALPHA	1
	Notes	12
2	THE INTERIM YEARS	15
	Notes	40
3	DESERT SHIELD	43
	Notes	76
4	DESERT STORM WEEK ONE	81
	Notes	158
5	DESERT STORM WEEKS TWO/THREE/FOUR	167
	Notes	188
6	DESERT STORM WEEK FIVE	191
	Notes	207
7	DESERT SABRE WEEK SIX	209
	Notes	241
8	POSTSCRIPT	245
	Notes	272
	APPENDIX	277
	ABBREVIATIONS	279

CONTENTS

	Page
BIBLIOGRAPHY	283
INDEX	291

Illustrations

Map

Week 1: 17–23 January	93
Weeks 2-4: 24 January–13 February	168
Week 5: 14–20 February	192
Week 6: 21–28 February	210

Photo

Brig Gen Dale Stovall	3
Capt Bennie Orrell	6
Jolly Greens were legendary in Southeast Asia	10
Brig Gen Rich Comer	30
Col George Gray	45
Capt Randy O'Boyle and Brig Gen Buster Glosson	54
The Apaches of Task Force Normandy	91
MH-53 pilot, Steve Otto	103
Crewmembers on the Stroke 65 recovery attempt	115
Col Dave Eberly, pilot of Corvette 03	123
Maj Tom Griffith of Corvette 03	123
The crew of Slate 46	144

Photo		Page
	1st Lt Randy Goff and Capt P. J. Johnson	148
	MH-53s stood CSAR alert at ArAr	176
	Capt Bill Andrews with other members of Mutt flight	205
	LTC Bill Bryan	230
	The crew of Bengal 15	234
	MAJ Rhonda Cornum	237
	Capt Tom Trask	262

Tables

1	JRCC Log	258
2	Analysis of combat losses	259

About the Author

Darrel D. Whitcomb, *right*, with Gen Charles A. "Chuck" Horner, USAF, retired.

Darrel D. Whitcomb is a 1969 graduate of the US Air Force Academy. He served three tours in Southeast Asia as a cargo pilot and forward air controller and subsequently flew the A-37 and A-10 with the 926th Fighter Wing and the 442d Fighter Wing. He also served in fighter plans on the Air Staff and in mobilization plans on the Joint Staff. Most recently, he served on the faculty at Air Command and Staff College and as the mobilization assistant to the commander of the Air Force Doctrine Center, both at Maxwell AFB, Alabama. He retired from the Air Force Reserve in 1999 with the rank of colonel. He was also a career airline pilot for Delta Airlines, retiring in 2003.

As a noted writer on aviation history and combat search and rescue, Colonel Whitcomb has published numerous articles in magazines and periodicals including *Air and Space Power Journal*. His first book, *The Rescue of Bat 21*, was published in 1998 (US Naval Institute Press). Still active in aviation, he recently completed a tour as a contract pilot in Iraq.

ABOUT THE AUTHOR

Colonel Whitcomb is a graduate of Squadron Officer School, Maxwell AFB, Alabama; Army Command and General Staff College, Fort Leavenworth, Kansas; and National War College, Fort McNair, Washington, DC. He lives with his wife in Fairfax, Virginia, where he works part-time as a contract analyst with TATE Incorporated.

Foreword

As a third-class cadet at the US Air Force Academy in July of 1970, I saw the war in Vietnam as a major determinant of my future. I just did not know then what part I might play in the conflict and felt a youthful, impatient need to decide on my Air Force career so I could see myself in what looked like the war of my generation. I found my calling on a beautiful California afternoon when 22 of us cadets were treated to a ride in an HH-53C Super Jolly Green Giant of the Air Rescue and Recovery Service, a part of the Military Airlift Command.

I enjoyed the ride. I sat beside a pararescueman, or "PJ," looking out at the treetops as we skimmed along. Turns were exciting, with the crew windows open up front and the wind flowing through the cabin. Unlike the rides I had taken in fighters and bombers, I was not breathing through a rubber mask, and I felt like I was in touch with the world below—a participant instead of a spectator. Afterwards, TSgt Stu Stanaland, the PJ, explained that the pilot, Maj Marty Donohue, was one of the best in the rescue business.

Months later back in school, we were all captivated by news of the attempt to rescue US prisoners of war (POW) at Son Tay in North Vietnam. When I learned that the raid had been conducted using rescue helicopters, I read all I could find about the mission. President Nixon later decorated the mission leaders, saying they had performed flawlessly and fought courageously without losing a man; and the mission had not succeeded only because of bad intelligence—the POWs had been moved from the Son Tay camp. As I watched the presentation on television, I spotted Major Donohue among the honorees.

Now I was hooked. I made a career decision to fly helicopters, get into rescue, and join those guys for life. That resolution held up through my senior year when the commandant of cadets told all us prospective helicopter pilots we were making a mistake that would deny us a rewarding career. "If you become helicopter pilots, you will not be in my Air Force," he said.

Still, it was what I wanted most to do and what I did. While in HH-53 transition, the most respected instructor pilot—an Air Force Cross recipient named Ben Orrell—told me rescue was a job to be proud of and that it would be sufficient in itself to make a career rewarding. Soon after, and still a second lieutenant, I flew my first combat mission—the *Mayaguez* recovery in 1975. We flew as wingman to another HH-53, which included Sergeant Stanaland among its crew members. I had joined those rescue guys.

The years between then and now have seen many changes but almost nothing new. The rescue men—and now also women—are called to enter dangerous territory, flying low and slow, going where a usually faster, less-vulnerable aircraft has recently been shot down. Instead of having a couple of days' time to prepare and entering the mission into the air tasking order (ATO) several days prior to execution, the rescue folks go there from a "cold start" launch from alert status. They could enhance their preparation by reading the ATO and plotting where the danger areas for shoot-downs might be, but a large, complex air campaign may cover an entire country. As you may deduce from this book, a major lesson learned from Desert Storm is that making the command and control arrangements murky and time-consuming further complicates the task of air rescue. Also, they may not be adequately equipped for the mission, yet the demand for immediate, must-succeed operations remains.

During the first Gulf War of 1991, air rescue was in a state of transition and largely performed by aircraft and crews assigned to special operations units. These units were equipped with helicopters originally procured to perform rescue and modified for their new missions. Remember, the Son Tay raid had been performed by crews and aircraft assigned to air rescue, but the mission concept more closely resembled a special operation. Likewise, the SS *Mayaguez* recovery was a rescue mission performed by air-assaulting a Cambodian island with US marines carried on Special Operations and Air Rescue Service H-53s. Special ops and rescue have closely related needs in terms of aircraft, crew training, and mission profiles. These are mission cousins, and the need for crews to consider themselves blood relations is clear.

FOREWORD

As detailed in this book, the young ones who aspired to become rescue guys in one generation often become the rescue leaders of the next. Just as I learned from and admired Ben Orrell, some of the pilots in the squadron I commanded in Desert Storm (Tom Trask, Tim Minish, Paul Harmon, and Mike Kingsley) commanded squadrons themselves in later conflicts in Kosovo, Afghanistan, and Iraq. As you read the stories of their missions in Desert Storm, you will also better understand their preparations to perform air rescue and special operations in those later conflicts. But those stories are for other authors.

In this book, Darrel Whitcomb provides a wonderfully objective accounting of combat search and rescue as we performed it during Desert Storm. He accurately discusses the decisions being made and their results. This work gives those who will do these missions in the future the chance to see things and ask the types of questions which, when answered, may lead to their success:

What is the command and control arrangement?

Who does the search part of search and rescue?

Who makes the launch decision?

What are the launch criteria for a rescue attempt?

Will these arrangements speed up or slow decision making?

Have we adequately rehearsed getting the launch decision passed to the alert crews?

Have we adequately rehearsed the combat link between the search aircraft and the rescue aircraft?

Do we have the right equipment in the hands of potential survivors?

Are the potential survivors trained for the role they must play in their rescue?

As Darrel Whitcomb examines these questions and much more in writing this book, we get the chance to learn and to

FOREWORD

become the rescue guys we would all like to be or would like to have standing alert in case we get into trouble. This book offers anyone a good start on that goal, on a wonderfully rewarding career, and a lifetime devoted to it.

RICHARD L. COMER
Major General, USAF, Retired

Introduction

"In combat operations, aircraft will be shot down," states Gen Charles A. "Chuck" Horner in his book, *Every Man a Tiger*.[1] Such has been the case since aerial combat became a form of warfare.

Implicit in this fact is the realization that the aircrews that fly those aircraft will be put at risk of capture. Over time, our nation has developed a robust capability to rescue airmen or any military personnel put at risk behind enemy lines. Each service has a rescue capability. The Air Force and Navy have dedicated rescue squadrons. The Army relies on its line helicopter battalions, as do the special operations units. The Marines have pre-trained teams, called TRAP (tactical recovery of aircraft and personnel), on call with each Marine air ground task force.

Originally, this mission was labeled search and rescue or SAR. SAR, in its original use by the US Air Force Air Rescue Service, meant recovery of personnel anywhere. When our rescue forces were deployed to Korea and Southeast Asia, they referred to all rescue missions, whether in enemy or friendly territory, by that term. The SAR stories from those conflicts are legion.

After Southeast Asia, analysts and force planners came to the realization that there was a fundamental difference between SAR in a permissive area and in an area that was not permissive (i.e., under enemy control). This second condition is now called combat search and rescue or CSAR. At the time of Desert Storm, the two forms of rescue were defined thusly:

> **Search and Rescue** (SAR): Use of aircraft, surface craft, submarines, personnel, and equipment to locate and recover personnel in distress on land or at sea.
>
> **Combat Search and Rescue** (CSAR): A specialized SAR task performed by rescue-capable forces to effect recovery of distressed personnel from hostile territory during contingency operations or wartime.[2]

The development of this rescue capability has been well established. Dr. Robert Futrell documented our efforts in Korea in *The United States Air Force in Korea, 1950–1953*. His work was followed by Dr. Earl Tilford's *Search and Rescue in South-*

INTRODUCTION

east Asia, which eloquently chronicled the heroic efforts of the rescue crews in that conflict who brought back literally thousands of airmen. It extensively documented what is now considered the "golden age" of rescue.

This work is meant to follow in those traditions and will focus on our CSAR efforts in the Persian Gulf War of 1991, or more specifically, the period of Operation Desert Storm, 17 January to 28 February 1991.

The initial reviews of CSAR in Desert Storm have been tepid at best. Dr. Richard Hallion, the Air Force historian, wrote in his *Storm over Iraq* that:

> One disappointment [in Desert Storm] was combat search and rescue (CSAR). . . . In the Gulf War, CSAR fell under the control of the Special Operations Command Central Command (SOCCENT), with all US CSAR forces placed under the management of Air Force Special Operations Command Central. While CSAR forces operated with their traditional reputation for bravery, dedication, and willingness to take chances to rescue downed aircrew, there were simply too few aircraft available to meet the requirements of both combat search and rescue and special operations needs. . . . Only three of 64 downed aircrew—roughly five percent—were actually picked up. . . . Indeed, only a total of seven CSAR missions were actually launched.[3]

Tom Clancy and Gen Chuck Horner, again in *Every Man a Tiger*, continue this theme when they state, "In DESERT STORM, the numbers rescued, as compared with the numbers downed, were low: eighteen men and one woman became prisoners of war as a result of aircraft shoot-downs. Seven combat search and rescue missions were launched, resulting in three saves. That's one saved for every six lost. Not an inspiring record."[4]

Initial impressions are powerful. The passage of time, however, allows a more thorough examination of the record as data becomes available and the veterans begin to speak. The examination of this conflict in a historical context suggests the following conclusions:

1. The Iraqi/Kuwaiti theater of operations was a challenging one for CSAR. The area was mostly barren, affording little cover for evaders to exploit or for aircraft to terrain mask. The weather was harsh, and the local populace was loyal to Saddam Hussein. Also, Iraqi air defenses were extensive and lethal.

2. This conflict reinforced the fundamental truth that the best CSAR strategy is not to get your aircraft shot down in the first place.

3. On the eve of Desert Storm, Air Force CSAR capability had been dramatically reduced from its peak during the war in Southeast Asia. Its primary recovery helicopters, the HH-53s, had been transferred to special operations. The remaining HH-3s were marginally combat-capable or survivable in a high-threat area. New HH-60s were only beginning to arrive. Navy combat rescue capability had been almost completely moved into its reserve component.

4. At the same time, the expectations of CSAR among the flying crews were high. This was based on the recorded history from Southeast Asia and the war stories passed down by the veterans of that conflict to the young troops in the Gulf. These expectations were reinforced by specific comments of some senior Air Force officers on the very eve of battle.

5. The non-rescue of an F-15E crew early in the conflict caused a morale problem among the F-15E units.

6. The Air Force component of Central Command (CENTAF) did not have a quick, accurate, and reliable way to locate downed airmen or discretely communicate with them.

7. Regardless, rescue personnel from the various services and Special Operations Command Central Command (SOCCENT) executed numerous joint missions and several SAR missions. Additionally, there were several intraservice rescues of note.

8. At no time were special operations forces (SOF) aircraft "not available" for rescue missions. On some occasions, though, SOF commanders did not dispatch aircraft because their assessment of the situation was that the mission was just too dangerous. It should be noted that during the conflict, zero rescue personnel were lost or killed on rescue missions directed by SOCCENT. However, a line Army crew was shot down on a rescue mission. That was Bengal 15, which was dispatched from the Air

Force tactical air control center. That particular mission will be examined later in detail.

9. Making CENTAF responsible for CSAR and then withholding operational control of the combat recovery assets was a violation of the principle of unity of command.

Two explanations are necessary to ensure the reader's understanding:

1. **Survivor call signs**. When an aircraft was shot down and the pilot ejected, he would maintain his call sign, like Benji 53. If the aircraft had a crew, then a discrete call sign had to be assigned to each individual. This was done by adding an alphabetical suffix based on the phonetic alphabet to the aircraft call sign. So, if an F-14 with a crew of two and a call sign of Slate 46 were shot down, the pilot would be referred to as Slate 46 Alpha, and the radar intercept officer as Slate 46 Bravo.

2. **CSAR as a mission**. CSAR is a three-phase process that involves: *locating* personnel at risk of capture in enemy territory, *authenticating* the personnel, and then prosecuting a *combat recovery*. This requires aircraft and assets of varying capabilities. Far too many historical tracts on rescue equate CSAR with combat recovery only. This lack of clarity was an issue in Desert Storm.

Overall, CSAR in Desert Storm appears to have been a mixed bag. Because of advances in precision weaponry, Global Positioning System (GPS) technology, countermeasures, and training, relatively few coalition aircraft were shot down. Forty-three coalition aircraft were lost in combat, most over high-threat areas. Eighty-seven coalition airmen, soldiers, sailors, and marines were isolated in enemy or neutral territory. Of that total, 48 were killed, one is still listed as missing, 24 were immediately captured, and 14 were exposed in enemy territory. Of those who survived, most landed in areas controlled by enemy troops. Of the few actually rescueable, six were not rescued for a variety of reasons, but primarily because of limitations in CENTAF's ability to locate them accurately and in a timely manner. Additionally, a lack of unity of command over all CSAR elements caused confusion and a strained working relationship between the joint rescue coordination center and SOCCENT, which operationally controlled the actual combat rescue assets. Regardless, the men who actually carried out the rescue missions displayed a bravery and élan so common to American rescue forces in earlier wars.

In this research effort I have many people to thank: Mr. Joe Caver and Mr. Archie Defante at the Air Force Historical Research Agency, Maxwell AFB, Alabama; Dr. Wayne Thompson at Air Force History; the officers and enlisted troops at the Air Force Rescue Coordination Center at Langley AFB, Virginia; Dr. James Partin in the Special Operations Command History Office at MacDill AFB, Florida; Mr. Herb Mason at the Air Force Special Operations Command History Office at Hurlburt AFB, Florida; and the librarians at the Joint Personnel Recovery Agency, Fort Belvoir, Virginia.

I also thank Col Gary Weikel, USAF, retired, Col Mark Bracich, and Lt Col John Blumentritt for patiently reviewing my manuscript and offering priceless constructive comments.

Finally, to all the veterans of this conflict, thank you for your time. Thank you for sharing your experiences with me. I have tried to relate your words as honestly and accurately as possible. For those of you who requested confidentiality, I have honored your desires. To those of you who refused to speak of the war, I respect that, too. War touches each man differently. Some men love it; some men hate it. All are changed irrevocably by the experience.

DARREL D. WHITCOMB
Colonel, USAFR, Retired

Notes

1. Tom Clancy with Gen Chuck Horner, *Every Man a Tiger* (New York: G. P. Putnam's Sons, 1999), 391.

2. See Appendix 6 to Annex C to USCINCCENT OPLAN 1002-90, 18 July 1990.

3. Richard P. Hallion, *Storm over Iraq* (Washington, DC: Smithsonian Institution Press, 1992), 246.

4. Clancy with Horner, *Every Man a Tiger*, 391.

Chapter 1

A Rich Heritage:
The Saga of Bengal 505 Alpha

We've got the best mission in Southeast Asia. I don't think that there's any larger pleasure than helping some guy out like that.

—Capt Dale Stovall

It was the 13th of April 1972. For three days the rescue forces from the 56th Special Operations Wing at Nakhon Phanom Air Base (NKP), Thailand, had been working to rescue an American who had been shot down near Tchepone, Laos, a central transshipment point on the Ho Chi Minh Trail. The two-man crew of an A-6 from Marine All-Weather Attack Squadron 224 stationed aboard the USS *Coral Sea* in the Gulf of Tonkin had ejected when their aircraft was hit by antiaircraft fire as it attacked a convoy of trucks. The pilot, Maj Clyde Smith (Bengal 505 Alpha), made contact with US aircraft overhead. His bombardier/navigator, 1st Lt Scott Ketchie (Bengal 505 Bravo), did not, and his fate has never been determined. Smith was alone, trapped along one of the most dangerous stretches of the road complex. To the planners and rescue crews at NKP, it was a dangerous tactical problem.[1]

But this was what US Air Force rescue crews did. Their units—equipped with a variety of helicopters and other support aircraft—had, by this late point in the war, made thousands of rescues. The process was almost routine, although each rescue provided unique challenges. Regardless, the men of rescue would fight to get him out. The nearest unit to Major Smith was the 40th Air Rescue and Recovery Squadron, also located at NKP. It was equipped with new HH-53 helicopters, which were specifically designed for combat recovery. Its crews were very familiar with the Ho Chi Minh Trail.

Like the Air Force, the US Navy also had rescue forces in Southeast Asia. Their primary rescue unit was Helicopter Squadron 7 (HC-7). With detachments aboard ships throughout the

Tonkin Gulf, its SH-3 helicopters and aircrews stood ready to recover downed crews, both at sea and on land.[2] They especially liked to pick up Navy and Marine aviators, but the location of Smith dictated that the 40th at NKP would get this mission.

For two days, the rescue planners and crews worked to get Smith out. Forward air controllers flying OV-10s and F-4s swept the area, first to locate the survivor and then to attack enemy guns massed nearby. Intelligence sources indicated that the enemy also had SA-2 surface-to-air missile (SAM) batteries in the area. A supreme threat to any rescue attempt, these were attacked by F-105s firing homing missiles and by several flights of aircraft with high-explosive bombs. To support the effort directly, the commanding officer of the *Coral Sea*, CAPT Bill Harris, launched 78 sorties to hit enemy forces.[3] The battle was unending as the Americans relentlessly attacked the North Vietnamese who, in turn, attempted to shoot down more of the orbiting aircraft. They knew that the Americans would try to rescue their man. They knew the Americans would send helicopters. They carefully husbanded their ammunition for the coming battle.

At the same time, North Vietnamese ground units began sweeping through the jungle looking for Smith. To impede their efforts, Air Force aircraft dropped "crowd control" weapons in the area. The powder that they spread made anyone who came in contact with it violently ill for about 30 minutes. This slowed the movement of the enemy troops and created a temporary barrier around the survivor.

Enemy defenses had to be subdued before a recovery helicopter could enter the area. Relentlessly, the forward air controllers directed flight after flight of Air Force, Navy, and Marine fighters as they pummeled the guns and SAM sites below. Several times the rescue commanders—flying in old Douglas A-1 Skyraiders and using the call sign Sandy—had considered and rejected sending in the vulnerable helicopters. The helicopter crews were ready to go, but they needed to have a reasonable chance of success.

Finally, on the afternoon of the 13th, the situation seemed to be propitious for a pickup attempt. As more forward air controllers and fighters struck the enemy positions around Smith, a large task force rendezvoused in the sky several miles west of his location. Leading the task force was an HC-130 aircraft

Photo courtesy of Dale Stovall

Brig Gen Dale Stovall learned the basics of combat search and rescue as a young captain flying helicopters in Southeast Asia, where he earned the Air Force Cross for one heroic rescue.

with the call sign King 27. Its crew would provide overall command and control and critical in-flight refueling to the rescue helicopters, if necessary, as routinely occurred on longer-range missions. The lumbering and vulnerable HC-130 would orbit well away from the recovery area.

The actual pickup would be made by Jolly Green 32, an HH-53 piloted by Capt Ben Orrell. The young captain was a colorful character. He grew up "out west" and had desired early on to be a bull rider. Discovering airplanes, he instead turned his quiet strength, courage, and resolve to rescue operations. Described by one contemporary as, "unpretentious, rock solid, steady, cool, and fearless," he had developed a keen sense of proportion, which seemed to give him the ability to quickly size up any situation.[4] He and his crew would fly as the "low bird," or the lead aircraft and would go in for Smith.

Backing them up would be another HH-53, Jolly Green 62, flown by Capt Dale Stovall. Dale was another star in the rescue community. A tall, lanky redhead, he was absolutely fearless and perpetually fought to fly missions. On his off time, he constantly read, researched, and probed for new ideas. To others, he appeared almost hyper and perhaps a bit of a bore. He was constantly out talking to the maintenance troops about equipment modifications or meeting with tactics officers from all manner of units to refine or improve operations. He lived rescue and thrived under the pressure. He and his crew would fly the "high bird" as backup for Orrell. Stovall was not happy. He wanted the lead position. But his time in Southeast Asia taught him that the situation below could rapidly change and push him to the fore. He knew his time was coming; if not today, then at some point in the future. The fighter pilots always needed the Jolly Greens.[5]

The two helicopters would be escorted by eight A-1s, led by Maj Jim Harding. Strong of body and quick of mind, Jim was the hard-charging commander of the 1st Special Operations Squadron (SOS), also located at NKP. His call sign was Sandy 01. In the coming battle, he would be the on-scene commander (OSC). He would have to make the difficult and dangerous decision to commit Orrell and his crew or, if necessary, Stovall and his. It weighed heavily on his mind.[6] Above them would be another armada of forward air controllers, fighters, SAM-suppression flights, and other supporting aircraft. It was an awesome flock of aircraft, there for one purpose: to get that American out of that mess.

The gaggle of helicopters and A-1s flew to a holding point west of Smith's position. From there the A-1s proceeded into the battle area while the helicopters held at high altitude. After trolling the area repeatedly, Major Harding felt that the enemy was quiescent and decided to call for Captain Orrell and Jolly Green 32. He directed two of his A-1s—Sandy 02 and Sandy

03—to acquire the helicopter and escort him in. Jolly Green 62 with Stovall stayed at the holding point as the backup.

Dutifully, the two A-1s headed off and joined with Jolly Green 32. Captain Orrell turned his aircraft and followed. As the helicopter and A-1s descended, an F-105 overhead detected a SAM site attempting to track the formation. The F-105 crew turned and fired an antiradiation missile at it. The deadly weapon exploded over the radar dish, completely destroying the fragile equipment and the men operating it. Jolly Green 32 and its escorts then spiraled down through an opening in the clouds. Once below, they were fully exposed to the enemy gunners waiting for them.

As they began to cross segments of the Ho Chi Minh Trail, several batteries of 23 millimeter (mm) and 37 mm guns began firing at them. As the tracers whizzed by, Orrell took evasive action. Spotting the enemy below, the gunners and pararescuemen, or "PJs," on board the helicopter began firing their miniguns. Sandy 02 kept firing white-smoke rockets ahead of the formation. As they exploded, their white plumes showed the pilots in Jolly Green 32 exactly where to fly. Sandy 03 darted back and forth around the helicopter, attacking enemy guns. Larger-caliber explosive shells flew back and forth as the three aircraft flew deeper into enemy-controlled territory.

Crossing a ridgeline, the helicopter was hit by 12.7 mm fire. One of the PJs, Sgt Bill Brinson, was slightly wounded but kept firing his minigun at the increasing number of enemy targets now all around them. Orrell pressed on to the waiting survivor.[7]

Visibility beneath the clouds was clear except for the billowing smoke from burning enemy wreckage and the air strikes directed by Harding. With the helicopter now approaching, he called for Smith to ignite a flare that would quickly produce a large cloud of red smoke. The smoke was easily identifiable and should reveal his exact location for Orrell. The helicopter pilot could then maneuver his aircraft directly over the survivor to make a pickup. Unfortunately, the smoke also revealed the survivor's location to any enemy troops in the area. This was one of the hazards of rescue work, but it had to be done.

Smith did as Harding instructed and lit his flare, but the smoke drifted into nearby trees and the helicopter crew did not see it. As Orrell settled his helicopter just above the trees and

Photo courtesy of Dale Stovall

Capt Bennie Orrell received the Air Force Cross, second only to the Medal of Honor in prestige, for the daring rescue of Bengal 505A, Maj Clyde Smith, USMC.

began to search, enemy troops sprang forth from every quarter and began firing at the vulnerable helicopter with weapons of various types. On board, crewmen in the back fired their miniguns in response. Above, Major Harding noted the enemy action and directed some of his A-1s to fire guns and rockets in close to the helicopter. He also called for other fighters to drop bombs and napalm on enemy forces a little farther away.

Hovering just above the foliage, the helicopter crew searched in vain for the survivor. Orrell was concerned because he knew that he now had the attention of every enemy gunner in that valley. He kept calling on the radio for Smith to give him directions. Harding directed Smith to ignite the other end of his flare. Orrell again reported that they could not see the survivor, so Smith decided to take action on his own and began running to an open area. Reaching the clearing, he saw the helicopter about 60 yards away and heading away from him. Smith screamed on his survival radio for the helicopter to turn around. Orrell reversed course and, almost immediately, his flight engineer spotted Smith. Orrell ordered his engineer to lower the jungle penetrator. When Smith saw the rescue device, he ran for his life as the totally exposed helicopter gently moved toward him.

Above, Harding and the other A-1 pilots knew that this was the critical moment. They formed an aerial "wheel" around the helicopter and laid down a withering field of fire on any enemy who dared contest the rescue.

Reaching the penetrator, Smith remembered the warning that survivors should let it touch the ground to discharge any static electricity before grabbing it. Static electricity was not high on his list of concerns at that moment. When the penetrator was still five feet above the ground, he tackled it. There was no shock. The device had a "horse-collar" and fold-down legs on which he could sit and secure himself. But he did not pull down the legs or wrap himself with the collar. Smith was still wearing his ejection harness, which had a metal clip that could be used to physically lock him to the hoist. He quickly snapped it to the connection ring and gave a "thumbs up" signal.

When the flight engineer, A1C Bill Liles, saw that Smith was attached, he began reeling the penetrator back on board. Orrell held his hover, and the A-1s continued to swarm above. As Smith reached the door, Liles pulled him in and then shouted for him to "get [expletive deleted] out of the way" as he swung his minigun back into place and resumed firing at the enemy below. He also told Orrell that the survivor was on board and cleared him to resume forward flight.[8]

Orrell relayed to Major Harding that Smith was on board. Harding directed him to "stay low" and egress the same way that he had come in. Orrell turned to the escape heading as the A-1s fell into a loose escort pattern for the flight out. Accelerating and climbing, the gaggle of aircraft took fire from several more enemy gun sites before clearing the area. The A-1s destroyed the sites with their remaining rockets and bombs.

The rest of the 90-minute flight back to NKP was uneventful. A huge reception awaited their return. As he climbed out of the helicopter, Smith was overwhelmed by the pride and professionalism exhibited by the rescue crews and truly humbled by the gratitude he felt for their selfless actions.[9]

After taking care of his injured crewmember, Captain Orrell inspected his helicopter. He found bullet holes throughout the aircraft. Several were in the cockpit area. Two rounds had gone through the main rotor blades, and one of the external fuel drop tanks was peppered with shrapnel from a larger-caliber gun. They were stark reminders of the vicious battle. But that was what he was paid to do.

Dale Stovall congratulated Orrell for an outstanding job. They debriefed the mission and began preparing for whatever was next. Soon, both were back on the flying schedule. The war went on.

For his actions that day, Capt Bennie Orrell was awarded the Air Force Cross, second only to the Medal of Honor in prestige.

Dale Stovall's moment came six weeks later when he was called to fly "low bird" for the recovery of Air Force Capt Roger Locher. Locher had been shot down over North Vietnam, near Hanoi. After evading for 23 days, he made contact with US aircraft. They alerted the rescue forces, and another armada was launched. It encountered combat equally as intense as the Smith recovery, but fighting through the enemy defenses, Stovall was able to recover Locher. It was an epic mission, for which Stovall was also awarded the Air Force Cross.[10]

Bennie Orrell and Dale Stovall were deeply affected by these missions. Their harrowing experiences solidified in both men a keen realization of what helicopters could and could not do when confronted by well-armed and alerted enemy forces. They would not forget these hard-learned lessons.

The medals awarded to Orrell and Stovall were not that uncommon among rescue forces. In fact the men of rescue were the most highly decorated group of airmen in that war. The reason was simple. As documented by historian Dr. Earl Tilford, the Air Force was responsible for the rescue of 3,883 personnel from all varieties of "at risk" situations during that long conflict.[11] In addition to combat rescues, this included troops who were medically evacuated or rescued from entrapped positions like the Citadel at Quang Tri in 1972.

This data is incomplete, for it does not include the thousands of US Army and Marine aircrews who were shot down and quickly recovered by sister ships and crews. Nor does it include the number of Navy or Air Force crews who were picked up by helicopters from other services. The stories are anecdotal, such as that told by former Air Force chief of staff, Gen Ronald Fogleman, retired, who as a young captain ejected from a flaming F-100 on 12 September 1968 and was immediately picked up by an Army AH-1 Cobra helicopter.[12]

In 1979 BioTechnology Inc. of Falls Church, Virginia, did a more focused study for the Naval Air Systems Command on combat losses. Specifically looking at what is now considered CSAR, the study indicated that the rescue forces of the Air Force and Navy rescued 778 downed aircrewmen throughout Southeast Asia.[13]

Included in the BioTechnology study was a note that in prosecuting those rescues, 109 aircraft of all types were shot down, and 76 rescue personnel were killed or captured. That meant that for every seven men saved, one rescue or support aircraft was lost, and for every 10 men saved, one rescue troop was lost.[14]

That study also included more startling numbers, which looked specifically at rescues within the high-threat areas of North Vietnam, where 125 aircrewmen were rescued. In the process, 41 rescue or support aircraft were lost for a rate of three saves per aircraft lost. More importantly, 34 personnel were lost for a rate of 3.7 saves per person lost.[15]

Those are sobering numbers. If the war in Southeast Asia taught one thing about rescue, it was that the helicopters—regardless of the amount of covering fire—were extremely vulnerable to enemy guns, missiles, and aircraft. Recognizing this, the men of rescue sought a better and safer way to conduct their mission.

Anecdotal evidence from several recoveries suggested that rescue operations under the cover of darkness could significantly reduce the danger of enemy defenses. Noting this, Air Force design engineers went to work to develop a night-recovery capability.

As in any engineering project, there were fits and starts, but by late 1972, the new system was ready for field-testing. Designated the Limited Night Rescue System, it combined a Doppler navigation system and both an approach and hover coupler. The prototype kits were shipped to NKP, Thailand, and installed on two 40th Air Rescue and Recovery Squadron HH-53 Jolly Greens. The aircraft were also modified with a low-light-level television system.

A RICH HERITAGE

Source: Darrel Whitcomb

The Jolly Greens were legendary in Southeast Asia, creating high expectations for combat search and rescue in Operation Desert Storm.

The crews immediately began training with the new technology. The test came on a dark night in late December when an AC-130 gunship, Spectre 17, was downed, again along the Ho Chi Minh Trail southeast of a deadly place called Saravane. Only two of the crew of 16 got out, and they were down along a heavily defended section of the road. It was a replay of Bengal 505.[16]

Jolly Green 32, an HH-53 modified with the Limited Night Rescue System, took off from NKP. The aircraft commander was Capt Jerry Shipman. Using the new technology, Shipman was able to fly straight to the first survivor and hover the big helicopter directly over him. Shipman sent one of his PJs down on the hoist. He easily found the man and brought him up. The system worked perfectly. Shipman then flew to the second survivor. He was in dense brush and took longer to recover. With both men on board, the crew of Jolly Green 32 hovered briefly to insure that no other crewmen were alive. Satisfied that they had the only survivors, they returned to NKP. Shipman was impressed. Even though they had been right along the dangerous road, they had encountered only minor ground fire.[17]

The tactical advantages of operating at night were obvious to all involved. Captain Shipman commented on the new system.

"The ability to go in there blacked out at night gives you a definite advantage," he understated.[18]

These rescue stories, as compelling as they are, were far too routine in our nation's costly war in Southeast Asia from 1964 to 1975. But they were part of combat, and every aircrew member who flew in the war held the hope that if he went down, the Jolly Green was coming. As one fighter pilot mused, "These Jolly Green boys are a breed all by their lonesome. As happy as we were to get picked up, the Jolly Greens were even happier to have done it."[19] Those feelings became part of the folklore of that conflict.

As the war ended, the conviction that whenever our aerial forces were dispatched to combat, the rescue forces would be there was imprinted on the minds of the combat aviators. For many, the extensive aerial bombing of Hanoi during Christmas of 1972 was actually seen as a huge rescue mission designed to free our POWs.[20]

But even that was not the end. In the spring of 1975, a joint US Air Force, Navy, and Marine task force was quickly assembled from the remnants of our forces still in the theater to rescue the crew of a hijacked freighter, the SS *Mayaguez*. The ship had been seized by Khmer troops off the southern coast of Cambodia, and intelligence believed that its crew was being held near the island of Koh Tang. Several large HH-53 and CH-53 helicopters were shot down by the massed guns and rockets of the enemy troops in a short and intense engagement. Coming almost as a footnote to the Vietnam War, this operation was little remembered except by the helicopter crews who knew that rescues in daylight visual conditions into areas controlled by massed enemy troops were no longer feasible.[21]

As one rescue troop later wrote, "[During] the last stages of the war in Southeast Asia conflict the enemy acquired more manpads [heat-seeking missiles] and better AAA. . . . The tragic losses during 'Mayaguez' seemed to mark the end of helicopters on a modern battlefield."[22]

The numbers of those rescued and those lost in rescue missions were sobering and pointed to a need to find a better way to do it. The Limited Night Rescue System looked promising. As our nation withdrew from and then moved beyond that war, the statistics and lessons hard learned got little notice. What endured instead were the memories of the heroism of the rescue crews. One A-1 pilot captured the essence of those memories when he wrote after one particularly brutal rescue effort, "The mission of the Sandy/Jolly team is to extract downed airmen from the clutches of the enemy. We have been reasonably successful in the past. Our proven capability to accomplish this mission has made a significant contribution to the high morale of the combat pilots in Southeast Asia."[23]

As the years passed—at places like Nellis Air Force Base, Nevada; Ramstein Air Base, Germany; Osan Air Base, South Korea; and hundreds of other bases and airfields—the veteran combat pilots of Southeast Asia passed those stories to a new generation of young tigers.

The helicopter pilots counted the patches on their aircraft and remembered the battles at places like Tchepone, Saravane, and Koh Tang.[24]

Notes

1. Lt Col Clyde Smith, USMC, retired, "That Others May Live," *Proceedings*, April 1996, 82–88.
2. CAPT Michael T. Fuqua, USN, "We Can Fix Combat SAR in the Navy," *Proceedings*, September 1997, 57.
3. Smith, "That Others May Live," 86.
4. Col Gary Weikel, USAF retired, e-mail to author, 20 May 2003.
5. Ibid.
6. Smith, "That Others May Live," 87.
7. Narrative of Capt Bennie Orrell, "Rescue of Bengal 505A," SEA Rescue and Recovery Files, file number 1118832, US Air Force Historical Research Agency (USAFHRA), Maxwell AFB, AL. The crew of Jolly Green 32 consisted of Capt Ben Orrell, aircraft commander; 2d Lt Jim Casey, copilot; A1C William Liles, flight engineer; A1C Ken Cakebread, PJ; and Sgt William Brinson, PJ.
8. Smith, "That Others May Live," 88.
9. Ibid.
10. Air Force Cross data supplied by Mr. Terry Aitken, USAF Museum, Wright-Patterson AFB, OH.
11. Earl H. Tilford Jr., *Search and Rescue in Southeast Asia, 1961–1975* (Washington, DC: Office of Air Force History, 1980), 155.
12. Chris Hobson, *Vietnam Air Losses* (Hinkley, UK: Midland Publishing, 2001), 162.
13. Martin G. Every, "Navy Combat Search and Rescue," BioTechnology Inc., Falls Church, VA, September 1979, 21.
14. Ibid.
15. Ibid.

16. Project Checo, Southeast Asia Report Search and Rescue Operations in SEA, 1 April 1972–30 June 1973, K717.0414-1, 48.

17. Capt Jerry Shipman, interview by the author, undated; and Hobson, *Vietnam Air Losses*, 244.

18. Shipman, interview.

19. Howard Sochurek, "Air Rescue Behind Enemy Lines," *National Geographic Magazine*, September 1968, 364.

20. John Darrell Sherwood, *Fast Movers: Jet Pilots and the Vietnam Experience* (New York: Free Press, 1999), xix.

21. See John F. Guilmartin, *A Short War, The Mayaguez and the Battle of Koh Tang* (College Station, TX: Texas A&M University Press, 1995), and Ralph Wetterhahn, *The Last Battle, The Mayaguez Incident and the End of the Vietnam War* (New York: Carroll and Graf Publishers, 2000) for excellent descriptions of this event. See also Col George Gray, interview by the author, 3 May 2001.

22. Lt Col Joe E. Tyner, *AF Rescue and AFSOF: Overcoming Past Rivalries for Combat Rescue Partnership Tomorrow*, National Defense Fellows Program, Headquarters United States Air Force and Air University, undated, 27.

23. Maj John Walcott, Sandy report, History of the 56th Special Operations Wing, January–March 1971, vol. 3, US AFHRA, Maxwell AFB, AL.

24. Shipman, interview.

Chapter 2

The Interim Years

They pulled all the HH-53s and crews down to Hurlburt overnight. [Eventually] they were permanently transferred to SOF.

—Brig Gen Dale Stovall

The war in Southeast Asia was a long and divisive one for the nation. During the lengthy period of withdrawal, aerial forces assigned primarily in Thailand covered the retreat of our soldiers and marines. The rescue units were among the last to leave. Personnel from the 3d Air Rescue and Recovery Group and the 40th Air Rescue and Recovery Squadron departed Thailand in January of 1976 on some of the last aircraft.

Returning to the United States, they were assigned to units flying the HH-53 and HH-3 that belonged to the Aerospace Rescue and Recovery Service (ARRS). It was part of the Military Airlift Command (MAC) and controlled all active and reserve component rescue units. Located at Scott AFB, Illinois, it also ran the collocated Air Force Rescue Coordination Center (AFRCC). This center was the clearinghouse for all continental rescues. It was authorized to coordinate with the Departments of Defense, Transportation, and Commerce; the National Aeronautics and Space Administration (NASA); and state and local authorities to direct and run civil search and rescue (SAR).

There was a large market for their services. Traditionally, Air Force rescue units stationed in the United States also supported civil SAR. Almost daily, the veterans of Southeast Asia were called upon to rescue lost hikers and foundering boaters or search for overdue aircraft. By 1981 the units of the ARRS had recorded their 20,000th rescue. Additionally, several of its members had received prestigious Cheney and McKay Trophy awards for some incredible recoveries.[1]

Refocused now on routine peacetime duties, little innovative thought was given to combat SAR, at least for the first few postwar years, but the value of the Limited Night Rescue System had been clearly recognized and had spawned some advocates. One advocate was a veteran of the 40th Air Rescue and Recovery Squadron. Finishing his tour at NKP in April 1973, Capt Jerry Shipman was transferred back to Scott AFB. The assignment was not happenstance. After the Spectre 17 rescue, his unit had been visited by MAC commander, Gen P. K. Carlton. The general was very proud of his rescue troops and took every opportunity to visit them. He had also been briefed on the Limited Night Rescue System and wanted to see it operate.

Captain Shipman was picked to take General Carlton on a night demonstration ride. They flew a mission into northeast Thailand and, using the Limited Night Rescue System, picked up some PJs. The general was impressed. At the debriefing, Shipman began to cover the strengths and weaknesses of the system. The general stopped him and, with a wave of his hand, acknowledged that the system did have some capability. Signaling for Shipman to be seated, he took the floor and proceeded to tell the assembled aircrews of other exciting technological developments which appeared to have utility in rescue. These were advancements in infrared night-vision capability and newer forms of precision navigation. Saying perhaps more than he should have, General Carlton shared exciting news of a new concept being developed at Wright-Patterson AFB, Ohio: a complete modification for the HH-53 called Pave Low.[2]

Pave Low was the successor to two earlier programs, Pave Star and Pave Imp. Both programs had attempted to develop a night and all-weather capable aircraft for worldwide SAR use, but both had been cancelled because of severe cost overruns. Pave Low was a new concept optimized from the mistakes of these programs and the promises of newer technologies.

The conversations with Shipman and other aircrewmen at NKP convinced General Carlton of the need. Returning to the States, the general threw his full support behind the Pave Low project and directed the ARRS to go forward with the proposed extensive modification for the big HH-53 helicopters to give them a full night and foul-weather rescue capability.

As conceived, it called for mounting a stabilized forward-looking infrared (FLIR) system, a new Doppler navigation system and computer, projected map display, terrain-following radar, and numerous improvements in self-protective capability.[3]

The Pave Low concept went through extensive conceptual design and modification before it was brought to fruition as the Pave Low III. The concept was shown to be feasible, and a prototype was modified in 1976. That aircraft went through an extensive operational test and evaluation that validated the concept. To fund the initial batch of modifications, General Carlton diverted funds from the C-5 fleet to modify the first 10 aircraft. The Air Force then dedicated funds to modify seven more aircraft and place them in the rescue fleet by 1980. This would give the ARRS a true day/night, all-weather rescue capability.[4]

As the development proceeded, the design engineers and officers in other tactical fields could not help but realize that the Pave Low had capabilities well beyond just combat recovery. Indeed, it possessed the ability to penetrate into high-threat areas for all manner of special operations at acceptable levels of risk.[5]

Holding a series of positions at ARRS headquarters, Jerry Shipman was intimately involved with the development of Pave Low and put his enthusiastic support behind it. Constantly seeking funding sources, he was a member of an ARRS briefing team that traveled to several Tactical Air Command (TAC) bases to sell the Pave Low concept for the modification of the CH-53s that TAC possessed. Almost identical to the HH-53, these aircraft had been used by Air Force special operations units in Southeast Asia to insert commando teams behind enemy lines. Some of them had participated in the *Mayaguez* mission.

Shipman and his teammates were convinced of the value of the Pave Low and zealous in their approach, but the reception, in most cases, was less than encouraging. They traveled to Bergstrom AFB, Texas, where TAC kept a unit of CH-53s and a numbered air force. Shipman and his team briefed the staff. The senior commander was unimpressed and only wanted to know when they were going to get those helicopters "the hell off of my ramp." They traveled to Langley AFB, Virginia, and briefed the TAC staff. Support was not forthcoming.[6]

In the end, though, it made no difference. The decisive and timely support of General Carlton ensured the success of the program.

The disdain that Shipman found in TAC for anything that had to do with helicopters reflected a larger distaste for special operations in general. During the Southeast Asia conflict, the Air Force special operations community had 550 aircraft of all types and more than 10,000 personnel. Since the war, the community had shrunk to 3,000 personnel with 28 aircraft. Most of these were assigned to the 1st Special Operations Wing (SOW) at Hurlburt Field, Florida, or Air Force Reserve and Air National Guard units. Smaller units were located in Okinawa and Japan. Their fleet consisted of old CH-3, UH-1, MC-130, and AC-130 aircraft, which, except for some of the C-130 variants, were more than 20 years old.[7] Within TAC, the special operations units were the lowest priority in funding.

Most amazingly, the heavy lift CH-53s that had done some great work inserting and extracting commando teams behind enemy lines in the war had been reassigned to a tactical communications unit. The special operations forces' MC-130s could insert the teams, but unless adequate airfields were available, they could not extract them. Helicopters were needed to guarantee that capability in all terrains.[8] The Pave Low was the answer, but TAC was not interested. Shipman found this disdain very frustrating.

Under MAC the ARRS commanded all Air Force rescue units worldwide. Those in the United States remained focused primarily on civil SAR, but some were also involved in significant actions outside the country. In 1976 rescue units participated in the emergency evacuation of Americans from Beirut, Lebanon. In 1979 another rescue task force was deployed aboard the USS *Saipan* for possible evacuation missions in Nicaragua when that country was taken over by communist forces.

That same year, Jerry Shipman, now a lieutenant colonel, was transferred to Kirtland AFB, New Mexico, where he took command of the 1551st Air Rescue and Recovery Squadron, the training squadron for the HH-53s, HH-3s, HC-130s, and the initial cadre for the Pave Lows. It was an exciting time, and Shipman loved his work. He felt that he held the future of rescue in his hands.

At about the same time, rescue elements were deployed to Turkey for possible use in Iran as that nation was swept by revolution. The men of the rescue task force were prepared to launch into Iran and conduct recovery operations for any Americans or other designated foreign nationals. Given the long distances involved and the turmoil sweeping Iran, these were assessed to be very high risk missions and were not flown. The rescue crews sat by helplessly as the bloody events took place and Iranian students seized the American Embassy and took 54 Americans hostage.[9]

In an effort to gain the release of the hostages, Pres. Jimmy Carter initiated several political and diplomatic initiatives. He also directed his secretary of defense to propose a military option. Given the lack of US forces or assets in the region, a conventional effort of any size would have taken quite a while to organize and deploy. Instead, the secretary decided upon a special operations effort, which was given the title of Operation Eagle Claw.

Several of the 1st SOW MC-130s and AC-130s were assigned to the mission. The helicopters of the 1st SOW were judged inadequate for the tasking because of the operational requirements. Instead, the planners decided to use RH-53 helicopters assigned to the Navy, for two reasons:

1. The helicopters would launch from an aircraft carrier, and the RH-53 was designed for that.
2. The RH-53s were also capable of carrying the heavy loads needed for the mission and operating in desert sand.

Initially, Navy pilots were selected to fly the helicopters, but they were replaced when several violated security rules and openly discussed mission details with family members. Evaluators also felt that the pilots lacked the tactical expertise for the mission. They were replaced by USMC pilots who also flew the aircraft.[10]

The rescue crews in Turkey were not pleased at what they were hearing. They were more than prepared to fly the mission with their HH-53s, regardless of the risk, yet they did not get the call. This would be an operation run by a select, hand-picked group of individuals drawn almost exclusively from the special operations community. It was completely compartmentalized, and they were not part of it.

The mission, initiated on 24 April 1980, was aborted when several of the helicopters broke down at an interim refueling site, Desert One. Then an RH-53 helicopter collided with an MC-130 aircraft as the helicopter tried to take off. The resulting explosion and fire were a horrible debacle.

The rescue crews in place in Turkey felt cheated. One crewman observed, "The Air Force Rescue Service is trained and practices and daily uses equipment that would have been far superior to what was used there. Why they were not used is hard to understand."[11]

The rescue forces had been the recovery element of choice at the Son Tay Prison in North Vietnam in 1970. That combined special operations and rescue force mission had also been a failure, but not because of the helicopters or aircrews. It had failed only because of a lack of timely intelligence. (The American POWs had been moved to a different location.) Had the Eagle Claw mission been delayed a few months, perhaps the highly modified Pave Low HH-53Hs finally coming out of the modification facility could have been dedicated to the operation.[12]

This failure showed a shocking lack of capability on the part of the special operations forces. They had no long-range helicopter capability. The hodgepodge combination of Navy aircraft with Marine pilots had failed. As one of the commanders of the ill-fated mission stated afterward, "You cannot take a few people from one unit, throw them in with some from another, give them someone else's equipment and hope to come up with a top-notch fighting outfit."[13]

Perhaps if TAC had kept its CH-53s in special operations units instead of putting them in communications units, they could have been the nucleus of such a force, especially if they had made the decision to modify them to the Pave Low configuration, but when Jerry Shipman and his team had tried to sell the concept, TAC had refused to do so.[14]

As US troops were being pulled out, unit commanders determined that some had possibly been left behind in Iran. An emergency order was sent to the 1551st Air Rescue and Recovery Squadron at Kirtland AFB to load two Pave Lows and crews on C-5s for immediate shipment to the Gulf so that they could be used in recovery operations for the stranded troops. Shipman and his troops quickly put together a package of aircraft and men and loaded them aboard the giant aircraft. The transports got as far as Dover AFB, Delaware, before the Eagle Claw commanders completed a detailed accounting of all troops and ascertained that their men were accounted for in full.[15]

Following the failure of Eagle Claw, President Carter ordered that a second operation be prepared. Code-named Honey Badger, this time the new Pave Low HH-53s would be included as part of a vastly increased armada of Army helicopters and the entire 1st SOW, but instead of including the rescue forces in the mission preparation, a new Joint Special Operations Task Force (JSOTF) under the command of the Joint Special Operations Command (JSOC) was formed.[16]

To support it, the chief of staff of the Air Force ordered the transfer of the Pave Lows from the Air Rescue and Recovery Service to the 20th Special Operations Squadron (SOS), a subunit of the 1st SOW at Hurlburt Field. This unit, equipped with HH-3s and UH-1Ns, had a long and distinguished record. Many of its crewmembers were veterans of Southeast Asia and had seen combat at places like Koh Tang and Tchepone. They were among the first in any operational unit to fly with night-vision goggles (NVG).

Lt Col Dale Stovall was amazed at how quickly the transfer happened. "[D]uring the Desert One mission, they had all the HH-53s down at Kirtland," he explained. "When . . . they lost the Marine birds, . . . they put JSOC together and started the second attempt. They pulled all the HH-53s and crews down to Hurlburt overnight. [Eventually] they were permanently transferred to special operations."[17]

At Kirtland AFB, Jerry Shipman was shocked by the turn of events. His aircraft and crews were taken from him with practically no explanation. Because of security concerns, he had not been "read in" on the Honey Badger operation. He knew that his troops were training hard for a special operation but did not know the details. Intuitively though, he was able to read between the lines and had a pretty good idea what was going on. His troops were involved in very high risk operations, and he had no control over any aspect of the mission. It made him very uncomfortable.

Shipman's concerns were realistic. As the Honey Badger armada grew, it began intensive training. At times as many as 40 helicopters would fly in mass formations with the pilots using NVGs. Accidents occurred, and an Army CH-47 and a Pave Low were lost. Four fliers were killed, but the preparations continued.[18]

Unfortunately, Honey Badger was never executed. After the failed Eagle Claw operation, the Iranian government divided the American hostages into several small groups and moved them constantly. Unsure of the hostages' location and facing unfavorable launch windows because of the short nights during summer months, US leaders were not willing to launch the second attempt until the fall. Eventually, diplomatic efforts led to the release of all personnel. Politically, President Carter paid a heavy price for the failed mission. He was defeated by Ronald Reagan in the 1980 election. The hostages were released as President Reagan took the oath of office.

The Pave Lows and crews stayed in the 20th SOS at Hurlburt. This abrupt series of moves stripped the ARRS of its best, most capable combat rescue asset. The ARRS still had a small fleet of HH-53s. Some did have the old Limited Night Rescue System; most were unmodified and their numbers were shrinking.[19]

To address this shortfall for the rescue community, the secretary of the Air Force, later that year, approved plans to purchase a fleet of modified HH-60D Nighthawk helicopters as the primary combat recovery asset. These aircraft would have some of the newer technology and arrive modified for night and all-weather operations. They would replace the aging HH-3s and HH-53s as they were taken for Pave Low modification. Except for a handful of aircraft, they would not begin to roll off the production lines until 1986.

In 1981, after a long hiatus, NASA resumed conducting manned space flights. In its mission statement, the ARRS was tasked to support NASA's space flights. As NASA's mission tempo increased, it put a heavy drain on ARRS resources and focused the command on noncombat operations. At the same time several after-action reports from Operation Eagle Claw and internal Air Force inspections suggested that, to increase emphasis on special operations forces, the Air Force should consolidate all of its helicopters under one organization. For unity of command, this organization would absorb both the ARRS and all Air Force special operations forces. The commander of the ARRS concurred and suggested that this would provide for force packages, which could then be placed at various locations around the world for both rescue and special operations tasking. The proposal was not acted upon that year.

The next year continued pressure from the National Command Authorities forced the Air Force to seriously consider the proposal. A combined Air Staff team from the Inspector General's office and the Office of the Deputy Chief of Staff for Plans and Operations conducted a thorough inspection and review of the Air Force's special operations capability. Their report confirmed earlier findings and recommended that all special operations forces and rescue forces should be consolidated into a single unit, preferably a numbered air force.[20]

Now assigned to the Air Staff, Lt Col Jerry Shipman did not concur with the report and tried to fight it. He feared that if rescue and special operations were merged, rescue would get sucked into the ebb and flow of special operations and eventually be subsumed by them.

The rescue forces required specialized equipment to operate effectively. Minor items such as modern survival radios for the aircrews, "Madden" survival kits designed to be dropped to survivors, and location devices such as the downed airman location system (DALS), all needed advocates. They had advocates in the rescue community. Shipman feared that in a merger, that advocacy would disappear.[21] His concerns were lost in the bureaucratic wrangling that swirled around the proposal. After a full Air Staff review, the chief of staff concurred with the studies and ordered the consolidation of the ARRS and the Air Force special operations forces under a newly activated numbered air force. After discussions between the commanders of TAC and MAC, the chief of staff directed that the new numbered air force would be assigned to MAC and would be the Twenty-third Air Force. The MAC commanders were happy with the arrangement because it gave them a way to protect their weather reconnaissance forces and rescue units in a numbered air force equal in stature to its Twenty-first and Twenty-second Air Forces that commanded the airlift units.[22]

Reporting to the Twenty-third Air Force would be the ARRS, which would still command all of the rescue forces, and a newly activated 2d Air Division located at Hurlburt Field, Florida, that would command all special operations forces units. Each entity would maintain its distinct identity.

This meant that Air Force special operations forces were being moved from TAC to MAC. To the airmen of special operations, this did not go down well at all. They had always felt as if they were the "lost children" of TAC, but at least there they were in a command that was focused on combat. MAC was the airlift command, and they were not happy.[23] Regardless, the transfer took place on 1 March 1983.

The first commander of the Twenty-third Air Force, Maj Gen William Mall, was pleased with the arrangement. Even though the two "communities" would maintain a separation, he felt that a natural synergy existed between them. This synergy had been exhibited many times, especially during combat operations like the Son Tay raid in Southeast Asia, the *Mayaguez* rescue operation, and numerous operations along the Ho Chi Minh Trail. In an interview with MAC's *Airlifter Magazine*, he stated:

> We created 23rd Air Force primarily to enhance the special operations forces (SOF) mission. The move capitalized on the synergism that exists between the special operations forces and the combat rescue forces because their mission, training, and equipment is similar. . . . The big payoff has been between special operations forces and combat rescue. Combat rescue has always augmented the special operations forces mission, but now we are training these forces in special operations tactics to a greater extent than ever before. Additionally, some special operations forces equipment is compatible and can serve both roles. The special operations forces Pave Low helicopters, for instance, have the capability to rescue a downed pilot in combat and still perform a special operations function without extensive modification of equipment or crew changes.[24]

As the consolidation took place, the Air Staff developed a long-term plan to acquire new helicopters to replace the aging fleet of HH-3s and disappearing HH-53s and upgrade existing equipment. During 1982 the ARRS received nine production-model UH-60A aircraft for rescue duty. Eventually upgraded to the HH-60G model and called the Pave Hawk, they were assigned to the 55th Air Rescue and Recovery Squadron and were also considered special operations forces–capable. This was the first of an extensive projected aircraft buy. Already, 60 HH-60Ds were programmed with a total project purchase of 243 aircraft. The HH-60Ds came equipped with an inertial navigation system (INS), terrain-following-and-avoidance radar, and FLIR, and reflected the advances in technology stimulated by the Pave Low program.

Money had also been programmed to modify 20 HC-130s to give them in-flight refueling capabilities for the helicopters. This designed buy was intended to satisfy the worldwide rescue needs of the Air Force. As part of this process, aircraft were designated for peacetime rescue and for wartime tasking. To differentiate the two, peacetime rescue was now formally called SAR, and wartime rescue was now called CSAR.[25]

Later that year, the Air Force Council on the Air Staff, upon reviewing the program, decided to cut the buy to 69 HH-60Ds and 86 HH-60Es. This was quite a reduction in capability. Not only were 88 aircraft removed from the purchase; in reality, the E-model aircraft was not nearly as capable as the D-model. The HH-60E was basically the same as the UH-60A. Then, as a final blow, Congress cut procurement funds for the program for another year in the FY 84 Appropriations Bill.[26]

Keying off the actions of Congress, the Air Force Council, in the spring of 1984, further scaled back the program from 155 total aircraft to 99 HH-60Ds. The Air Staff then reduced these 99 HH-60Ds to 90 HH-60As. Their reason was financial—the HH-60D cost $22 million; the HH-60A cost $10 million, but it was a much less capable aircraft. Even this was lost the next year when the Air Force Council again reviewed the program and decided to cancel the buy altogether.

Col Tony Burshnick was at the meeting when that decision was made. Serving at that time as the chief of plans for MAC, he attended the council meeting with several younger staff officers and recalled the discussion:

> Our case was being presented by a rescue guy from the Air Staff. . . . The Vice-Chief [of the Air Force] . . . listened to this pitch and he said, "That [HH-60] is a great, great helicopter." And then, of course, the price tag came up. [The board members] yakked about it around the room and they finally decided that they were going to kill it. It was too expensive. I said, "Wait a minute. You're killing rescue service." And the guy said, "If we put all that money into the H-60, there won't be any money to buy fighters so there won't be any fighter pilots to rescue." . . . So there was no [HH-60].[27]

The commander of MAC, Gen Thomas Ryan, challenged this decision, but his priority at that time was the new C-17 heavy lift cargo aircraft, and his efforts on behalf of the ARRS were to no avail. His staff began work on another rescue helicopter master plan, but at this point, there was no money and little support for new helicopters for rescue.

At the same time, the special operations forces were also facing some danger. In November 1984, the Army and Air Force signed an agreement called the "Thirty-one Initiatives." This agreement was designed to resolve certain long-standing disagreements between the services on a wide range of issues. Initiative 17, as agreed to, called for the transfer of the special operations heavy lift mission to the Army. One option considered was to transfer the Pave Lows to that service.

The Air Force and Army began to work through the agreement. Another option called for the Army to perform the mission with heavily modified MH-47 helicopters. After three aircraft accidents, the Army realized that whatever aircraft was selected for this mission would have a huge operational and logistical tail and, to be properly utilized, would have to have direct access to a tanker fleet of HC-130s. They began to reconsider Initiative 17.[28]

The chiefs of the Air Force and Army held a meeting to review the entire initiative. After a spirited discussion, they agreed to suspend the initiative and leave the Pave Lows with the Air Force.

That decision was reaffirmed by Congress when it passed legislation directing the Air Force to keep the long-range helicopter support mission instead of giving it to the Army for both combat recovery and special operations missions. Additionally, the Air Force was given a green light to modify the remaining rescue HH-53s and the TAC CH-53s to the enhanced Pave Low III configuration now called the MH-53J.[29]

Above the services, there was another wind blowing. Powerful members in Congress—upset with the recurring problems in the military, as exhibited by the failure of Operation Eagle Claw and subsequent operations in Grenada and Lebanon—decided that the services needed to be fundamentally reorganized. Perhaps the solution was to create a new special operations force or command as a new "sixth service." One of the leading proponents of this was Cong. Dan Daniel (D-VA), chairman of the powerful House Armed Services Committee (HASC). In August 1985 he stated, "the current administration has been pursuing the revitalization of our SOF capability. The Secretary of Defense has assigned the highest priority to this effort. Congressional support has been strong (and is growing); media attention has been intense (and generally favorable); and the public interest is intensifying."[30]

The new commander of the Twenty-third Air Force, Maj Gen Robert Patterson, also sensed the winds of change. He directed his staff to do a study called "Forward Look." Following Congressman Daniel's lead, that study recommended quickly growing the Air Force special operations forces from one wing-equivalent to three. To do so without any aircraft procurement, they would have to take aircraft from the rescue forces, further weakening ARRS's capability to perform combat recovery.

General Patterson briefed the plan to the new MAC commander, Gen Duane Cassidy. He approved it, as did the Air Force Council, in May 1986. The plan would direct the transfer of all of ARRS's HC-130 tankers and remaining HH-53s to the Air Force special operations forces for the creation of overseas wings.

Subsequently, the 2d Air Division staff did an assessment of combat rescue capability after the transfer of assets. Their finding stated that "all the assets remaining in rescue units were non-combat capable." It continued on to say that special operations forces could "conduct combat rescue on a relative priority basis. . . . [D]owned aircrews and other personnel requiring recovery must plan to survive and evade until air forces can come and get them."[31] In fact, during this period when all of our warfighting planning was focused on the big European war, combat rescue was almost not discussed. The reason was simple: the threat was so overwhelming that planners did not feel that the meager rescue forces would be able to survive penetrating the projected heavy Soviet forces. Hence the aircrews were told that if they were downed, they were to evade and move to certain specific locations. At predesignated times, special operations helicopters would fly by and pick up any crews there.[32]

And it seemed, at least from the tactical unit perspective, that CSAR skills were allowed to atrophy. CSAR was not practiced at the "Red Flag" exercises that took place constantly at Nellis AFB, Nevada, nor at Fort Irwin, California, where tactical aircraft deployed on two-week stints to fly close air support with Army units. There was no formal training for A-7 or A-10 pilots to develop skills as "Sandy" on-scene commanders to tactically direct rescue operations, as so many A-1 pilots had done in Southeast Asia.

At the same time, though, the men in the HH-3–equipped rescue squadrons trained earnestly for combat recovery. They took part in exercises when they could and suffered losses in terrible accidents. The instructors at the HH-3 school at Kirtland AFB, New Mexico, pushed their students and demanded near perfection. Regardless of what the "official" studies concluded, they considered themselves ready for combat with what they had and fully intended to do so. They were the men of rescue. If their services were needed, they expected to go.[33] They did not intend to be left out of the action as the rescue forces had been in Iran.

Working with Congressman Daniel in Washington was an aggressive staffer on the HASC, Ted Lunger. Ted was a former Army Green Beret and "A" team leader and operations officer with the CIA. He had served several tours in Vietnam and Lebanon and moved to the legislative branch of government in the late '70s for "family reasons." There, he quickly developed a solid reputation as someone who could get things done. Congressman Daniel relied on him to oversee issues related to special operations and rescue.

Working quietly behind the scenes, Lunger carefully studied the history of special operations and rescue in Southeast Asia and the Desert One and Grenada operations and concluded that both mission areas suffered from benign neglect. He noted that, "The parent services will not resource the mission if left to themselves. They won't fly the mission, they will not package it. They won't train for it. They won't spend the money. The closest equivalent you had in the AF in peacetime was the [ARRS]. I traveled around and studied what the [ARRS] did in peacetime and there wasn't any training. . . . Now, were these guys doing wonderful things with minimal resources? You bet . . . ."[34]

Based on his findings, Lunger crafted a proposal to create a National Special Operations Agency (NSOA) similar in scope and authority to the CIA. It would include components from all of the services. In the proposal, the NSOA would be responsible for maintaining a CSAR capability, although he labeled it theater SAR, or TSAR, because it would belong to a theater commander in chief. The Air Force component would consist of three wings. Lunger explained:

> In the NSOA legislation, theater SAR (TSAR) was a special operations forces mission under the law. TSAR was the same as CSAR. What the deal was—the CINC was supposed to get an augmented special operations forces package, a combined Wing, out in each of the unified commands under the legislation. In each, there was a heavy rescue component because we were dual tasking them for the insertion and exfiltration for the special operations forces missions. The CINC was given the capability and assets and was given command of the assets under that concept. . . . Under the NSOA, the theater CINC was to have his own heavied [sic] up special operations forces wing. Inherent in that wing was a squadron-plus of Paves. The follow on units would be MH-60s.[35]

The House of Representatives passed his proposal as part of the 1987 Defense Authorization Bill.

This alarmed Senators Sam Nunn and Bill Cohen, who had been instrumental in passing the Goldwater-Nichols Act in 1986. That landmark law had given commanders of unified and specified commands operational control over all assigned combat forces from any service. In reviewing the NSOA concept, they felt that the agency would be too easily manipulated by the various service chiefs. Instead, they wanted to create a separate unified command for special operations with a commander who had his own command and budget authority separate from the service chiefs. In a series of meetings, they and Congressman Daniel were able to craft legislation creating the Special Operations Command (SOCOM) as an amendment to Goldwater-Nichols.

Working behind the scenes, Ted Lunger pushed hard to make the force as strong as possible. He wanted to include the principles behind the NSOA concept in the SOCOM legislation, especially the TSAR forces. Working closely with Mr. Jim Locher, the assistant secretary of defense for Special Operations/Low Intensity Conflict, Lunger was able to incorporate most of his proposals into the legislation.[36]

That summer Congress passed, and President Reagan signed, the 1987 Defense Authorization Act, which included language creating the US Special Operations Command. TSAR was listed as the ninth of nine primary missions.

As a unified combatant command, SOCOM would have components from each of the services. The Air Force portion would

THE INTERIM YEARS

Photo courtesy of Rich Comer

Brig Gen Rich Comer was a young lieutenant when he flew on the *Mayaguez* rescue in 1975. He followed the rescue mission when it was moved to the special operations forces between Vietnam and Desert Storm.

be the Twenty-third Air Force. It would be comprised of three Special Operations Wings (SOW): the 1st at Hurlburt Field, Florida, the 39th in Europe, and the 353d in the Pacific. To align his command to the new reality, General Patterson moved the Twenty-third Air Force to Hurlburt Field and deactivated the 2d Air Division. Its special operations units immediately began to work and train with the other elements of SOCOM.

The effect of all of this on the rescue community was disheartening. Except for one unit in the Pacific, they had been stripped of their most capable combat rescue helicopters and tankers and had shrunk to one wing-equivalent in total force structure. The assets taken away were being used to build up the three special operations wings. What aircraft remained were not considered combat-capable.[37]

More importantly, their owning command had completely changed its character—they were orphans in a command with a much different focus. And most importantly, their future, as represented by the procurement program for new helicopters, had been cancelled. Of course, the troops did not miss the changes. They began to transfer over to special operations.

Administratively, the Twenty-third Air Force was still part of MAC and still owned distinctly non-special operations forces elements like weather reconnaissance and aeromedical airlift units along with the ARRS. The first commander of SOCOM, Gen James Lindsey, was unhappy with that arrangement. He wanted nothing to do with these units and asked the Air Force chief of staff, Gen Larry Welch, to transfer them back to MAC and to redesignate the Twenty-third as a major command, as it had been when originally transferred to SOCOM. General Welch agreed. The non-special forces units were transferred back to MAC, and the Twenty-third Air Force was slated to become the Air Force Special Operations Command (AFSOC) in May of 1990.[38]

At about the same time that the Twenty-third Air Force was moving to Hurlburt, a stocky, slow-talking young major was reporting to McClellan AFB, California, for duty as an HH-53 rescue pilot. Maj Rich Comer was an old hand at rescue. As a young lieutenant, he had served in the 40th Air Rescue and Recovery Squadron at NKP and was there in 1975 when they flew in the *Mayaguez* effort. On that fateful day, he was copilot on one of the HH-53s that went back in to rescue one of the crews feared left behind. They searched along the beaches until they were notified that all personnel had been accounted for. Comer could still hear the sound of the enemy rounds ripping through his lumbering helicopter.

Arriving in June from an assignment on the faculty at the Air Force Academy, where he had shown great promise as an academician, he began his HH-53 checkout only to learn in September that the unit would be closing and all aircraft would be sent to Pave Low conversion. Comer was offered a job flying the Pave Low at Hurlburt in the 20th SOS. The squadron commander, Lt Col Bo Johnson, personally offered him the job because he needed a few rising field-grade officers with some upward potential. Comer took it and became a special operations pilot. He had already done some work with them and was intrigued by the variety of missions they flew.[39]

Starting his Pave Low transition, Comer found the flying to be very challenging, especially the night pattern work while wearing the night-vision goggles. Some of his training sorties were deeply frustrating, as he struggled to learn the "special ops way" of doing things.

Comer saw immediately the difficulties of combining crews in the squadron from two different backgrounds. There was definitely animosity between the rescue guys and the special operations guys. He was struck by the arrogance of the special operations pilots and could feel their disdain for the rescue "pukes." The special operators had high performance standards and did not feel that the rescue crews met them. Transferring in, the rescue guys had a choice: they could pick either the "special ops way or the highway."[40] Several of the rescue guys could not handle the flying and did ask to leave.

This process of blending the crews became Comer's task. Bo Johnson had specifically hired him for his maturity and organizational leadership skills, and Comer worked closely with all the crews to form them into one cohesive group as the unit rapidly expanded to 20 aircraft and 30 fully formed crews.[41]

AFSOC needed the transferees because it was growing rapidly in an ever more dangerous world. Many of the "old head" rescue guys did make the transition, but just as many decided that they had seen enough, had enough, whatever. They transferred back to what was left of rescue or retired. To many, it was the end of an era.

Gary Weikel, then a lieutenant colonel and a former rescue pilot, made the transition into Pave Lows with the 20th SOS. He described what it meant to be assigned to the unit during this difficult transition when he wrote, "It became . . . clear if you came to fly Pave Lows in the 20th SOS, be prepared to fly all night, scare yourself . . . on innumerable occasions, live on the road in absolutely [expletive deleted] conditions, go on a moment's notice and be unable to contact your family for weeks/months on end, and be satisfied that you were doing something good for the country that no one else could do and you were part of that 'band of brothers' that would fly through the gates of hell for/with you."[42]

With its creation SOCOM was almost immediately involved in operational missions that involved direct action, unconventional warfare, and antiterrorism actions. Its assigned forces, including those from the Air Force, were quickly engaged in Operation Earnest Will in the Persian Gulf from 1987 to 1989.

This intense operation involved protecting oil tankers during the Iran-Iraq War. In late 1989 they were thrust into Operation Just Cause in Panama, which tasked them to the limit. With real-world concerns and operations, little time or effort was spent on TSAR.[43]

In fact, just the opposite was occurring. Senior planners at SOCOM realized what a tremendous capability they had in the Pave Low. It finally gave them the ability to clandestinely take their troops deep into enemy territory. They spent vast sums of money to give the aircraft all of the latest technological advances such as satellite-based communications and navigational devices. The idea of solely dedicating these aircraft to rescue or of having them sit idle on rescue alert was anathema to them. It just had too many other capabilities that needed to be utilized on an almost daily basis.[44]

Then, in late 1988, the Air Staff directed that combat recovery would not be included in the designed operational capability (DOC) statement of special operations units. This formal step severed any official connection between the special operations forces units and combat recovery.[45]

Buster C. Glosson, then an Air Force colonel serving in the legislative liaison division of the Air Staff, watched this process from the inside and was dismayed by the turn of events. He said, "I was appalled at the situation with CSAR prior to [Desert Shield/Storm]. . . . I thought that the Air Force leadership had been shortsighted. . . . The decision made by the corporate Air Force was to spend the absolute minimum on SAR or CSAR, and we would leave that responsibility to somebody else."[46] In the not-too-distant future, he would have to deal directly with the results of those actions and decisions.

Several events occurred that convinced Air Force leaders that they needed to revitalize the service's combat recovery capability. First, the commander of the Pacific Air Forces (PACAF) noted that two rescue units assigned to his command were slated to be converted to special operations units. Effectively, this would leave him with no combat rescue capability in his immense theater. To fix this he pushed for a rejuvenated and enhanced rescue service.[47]

At almost the same time, the Air Staff published a Rescue Force Structure Plan (RFSP). Almost two years in the making and reflecting the monumental changes to the rescue forces throughout the decade, this document laid out a comprehensive program to rebuild an Air Force combat rescue capability. Its stated goal was the recovery of 65 percent of all aircrews downed in combat. Reflecting the new realities of command and control dictated by the Goldwater-Nichols Act, the plan directed that the Air Force would retain operational control of its recovery helicopters through a rescue coordination center (RCC) and discussed both preplanned and immediate response missions in a variety of threat situations.

In a classified section, the plan addressed shortfalls and suggested equipment necessary to achieve the goal of the plan and a timetable to do so. Signed by the vice chief of staff of the Air Force, the plan was given wide dissemination throughout the rescue community.[48]

In response to these events the MAC commander, General Cassidy, resolved to take action on two fronts. First, he wanted to fully reclaim the ARRS. He had his staff develop a plan to have it revitalized, rebuilt, and in place by 1995. He briefed it to the senior Air Force leadership and was directed to have it up and running by August 1989. Additionally, the leadership agreed to remove the ARRS from the disbanding Twenty-third Air Force, rename it the Air Rescue Service (ARS), and relocate it to McClellan AFB, California. From there, it would, like the ARRS, command and control all active duty air rescue units and the Air Force Rescue Coordination Center, still at Scott AFB, Illinois. It would also serve as the gaining command for all reserve component rescue units upon call-up.

The ARS was activated at McClellan on 8 August 1989 and assigned directly to MAC. Upon activation, the rescue forces and special operations forces were separate. Since the war in Southeast Asia, the rescue community had come full circle. Unfortunately, in making the journey, it had lost its best combat rescue aircraft. It would take time to correct the shortfall.[49]

To address the aircraft issue, General Cassidy revitalized the old initiative to procure new helicopters. This effort began to pay off when the Air Staff programmed money to buy 16 UH-60As, which could then be modified for rescue duty and fielded as HH-60Gs, which were technologically upgraded HH-60Ds. These aircraft were to be delivered by 1989 as the first part of a steady buy of 10 aircraft a year for several years. They would be used to equip active, Guard, and Reserve units.

With such strong support, the newly reconstituted ARS set out to rebuild. It took control of HH-3–equipped rescue squadrons in Japan and Korea that had been slated to transfer to the 353d SOW, as well as squadrons in Alaska and Iceland, and designated them as combat rescue units. In February 1990, the ARS began to receive its first HH-60Gs. They were assigned to the units in Korea and the Air National Guard. Additionally, the ARS made plans to stand up a rescue squadron equipped with HH-60Gs at Nellis AFB, Nevada. It would be the "first combat mobility unit based in the continental United States" and would activate on 1 January 1991.[50]

Upon activation, the ARS published its mission statement. It stated, in part:

> Air Rescue Service (ARS) is the focal point for USAF rescue. . . . The missions of ARS include combat rescue, peacetime SAR, humanitarian SAR, support for the National Aeronautics and Space Administration, and worldwide USAF Rescue Coordination Center activity. . . . The primary mission of ARS is combat rescue which traditionally involves the helicopter recovery of downed aircrew members from a hostile environment, usually supported by HC-130 tankers and dedicated fighter aircraft.[51]

At that point, the staple aircraft of the ARS units was still the venerable HH-3. Since the remaining HH-53s had been transferred for the Pave Low modification, the HH-3s and a few UH-1Ns were the only helicopters available until the new HH-60s arrived.

Maj Mark Tucker was in a unit as a combat-ready HH-53 pilot when the aircraft were taken away, and he had to convert to the HH-3. He was appalled because the HH-3s were not capable of operating in a high-threat environment. He remembered that "[the HH-3] did not have any defensive systems on it. There was no radar to allow it to fly low level. There were no threat defensive systems like we had on the '53. It was all 'plan the mission and go fly.' The first you would know of a threat was when the rounds started impacting the aircraft or visual identification of a missile launch. So crews trained to that, but it was all based on 'Hey, anybody ever seen this?'"[52]

Some HH-3 units did have flare-dispenser units on their aircraft, but it was not much when considering all the modern threats that HH-3s would have to face in a high-threat conflict. Considering this, the HH-3s could practically be used only in low- or no-threat environments.[53] Regardless, the men in the units trained diligently and were ready if called.

There was one more issue that needed to be resolved with the transfer of assets from the ARRS/Twenty-third Air Force to ARS. Under the Twenty-third, all Air Force pararescuemen had been assigned to the squadrons belonging to the 1720th Special Tactics Group (STG). With the activation of the ARS, one squadron, the 1730th Pararescue Squadron, would be split out and transferred to McClellan AFB, where it would collocate with the ARS. Detachments of PJs from the 1730th would then be assigned to the various rescue units around the world. The 1720th STG would remain at Hurlburt as part of AFSOC.[54]

Navy

The US Navy also maintained rescue units, which carried a long and rich heritage written, in part, by the men of HC-7 who flew SH-3s off ships in the Tonkin Gulf. Its crews flew countless recovery missions for flyers down in the waters of the Gulf or in North Vietnam.

After the war, the Navy deactivated that unit and transferred the aircraft and mission to HC-9, a Naval Reserve unit. Many of the HC-7 personnel also transferred over, and they kept the corporate CSAR memory alive in the Navy.

Still flying the SH-3, HC-9 maintained a small but well-motivated cadre until 1989 when it was split into two new reserve units, HCS-5, based at NAS Point Mugu, California, and HCS-4, based at NAS Norfolk, Virginia. Both squadrons could perform combat rescue and naval special warfare operations, and each was assigned eight new HH-60H Seahawks. Every aircraft was equipped with the latest in navigation gear, communications, and armament, and was a huge improvement over the older aircraft it replaced.[55] Additionally, all helicopter units that deployed with the fleet had the ability to assume recovery alert if needed on an ad hoc basis.

Survival Radios

Since the Korean War, the US military had recognized the value of equipping airmen with survival radios. These small, handheld devices served two purposes:

1. They allowed the downed personnel to make voice contact with covering and rescue forces to facilitate rescue.
2. They also had a "beacon" mode that sent a loud, piercing wail that could be homed in on by rescue forces, providing another way to find survivors.

These radios were key to the recovery of hundreds of crewmembers from the jungles of Vietnam. In the later years of the war, all crewmembers carried two URC-64 radios. These were reliable four-channel radios. Most flyers carried extra batteries.

To home in on the signal quickly and accurately, the Air Force had developed an Electronic Location Finder (ELF). This device could pick up the survival radio signal and give the helicopter crew accurate guidance to the survivor, but it did not have a way of measuring the distance to the man. Hard experience in Southeast Asia highlighted the need for this capability.[56]

After the war, as the URC-64s wore out, the Air Force and Navy replaced them with the PRC-90. The new radio had similar capabilities, but broadcast on only two frequencies.[57] Like the URC-64, it could guide helicopters using the ELF.

The radios broadcast on well-known international frequencies, and over time, the tactics and techniques of rescue became common knowledge. Any potential adversary of the United States could easily figure out how to exploit them—either broadcast false signals or home in on the signals themselves.

By the time of Desert Storm, a new radio had been designed for the aircrews, the PRC-112. This radio had the ability to transmit on three common international frequencies and two programmable frequencies. It also had a new feature built into it—a discrete capability to precisely guide an aircraft to it if the aircraft had been equipped with a homing device called the Downed Airman Location System or DALS. This was a vast improvement over the ELF.

The Navy HH-60s had this homing capability, as did the AFSOC MH-53s. None of the residual Air Force helicopters had it.

THE INTERIM YEARS

Approaching the conflict, the Navy had begun to replace its PRC-90 radios with PRC-112s, as had the Special Operations Command. The Air Force had not, although the Rescue Force Structure Plan had clearly recognized the need to do this.[58] The radios cost about $3,000 each, and TAC had chosen not to spend the money.[59] It would send its crews into combat in any near-term conflict with radios easily exploitable by the enemy.[60] And each crewmember was issued only one radio.[61]

Understanding the importance of quickly locating downed airmen, the Air Force had programmed the ability to listen for and locate any emergency calls into several of its intelligence assets. Each asset, however, had an error factor in it, an "error probable" in military terms. This meant that a radio might be located, but instead of a precise fix, the location report would say "location is within an X-mile circle of Y latitude and Z longitude." For example, one of the assets for locating survivors was the Search and Rescue Satellites (SARSAT) system. This constellation of satellites in polar orbit could quickly pick up any emergency signals, but its "error probable" was 20 km or about 12 miles.[62]

The rescue forces needed a more accurate position than this to commit for a recovery, especially in a high-threat area. In Southeast Asia, traditionally, either forward air controllers or fighters would quickly pinpoint a downed aircraft's location. If this were not possible, a flight of A-1 Skyraiders would roar into the area and use homing techniques to locate the survivor before committing a vulnerable helicopter for the pickup. This was dangerous business. During that war, 191 Air Force A-1s were lost to enemy guns and missiles, most on rescue missions. Nine HH-53s met the same fate.[63] To operate in high-threat areas, fighter-type aircraft would be needed to perform this function, but neither the Air Force nor the Navy modified any fixed-wing aircraft with the DALS.

The Air Force planned to use other assets instead to locate its survivors. The airborne warning and control system (AWACS) aircraft with their excellent radar and intelligence capabilities and the "Rivet Joint" RC-135 electronic collection aircraft appeared to have the ability to locate downed survivors. The RC-135s and AWACS aircraft had been equipped with global positioning systems (GPS) for accurate navigation. These devices received position data from a constellation of satellites 11,000 miles above the earth. The accuracy of the system allowed the aircraft to precisely determine its position within 10 meters. Using the latest map reference datum, called Worldwide Georeferencing System 1984 (WGS-84), they expected to be able to accurately plot any form of radio transmitter. Since the AWACS and Rivet Joint aircraft would be present in any theater of operations, they could provide the location capabilities for downed airmen. This system, however, had not been tested in the rigors of combat.[64]

The development of the GPS was an exciting advancement. Through the ages warriors had wrestled with the problem of accurately determining position and movement. The highly accurate GPS seemed to offer a solution. All US military services were rapidly moving to incorporate the system in their forces. Already, 12 satellites were in orbit and operating. When fully operational the system would consist of 24 satellites that would provide 24-hour navigational coverage around the globe. Besides the RC-135s and AWACS, receiver kits had already been installed on the AFSOC MH-53s and Block 40 F-16C/Ds. The Army and Marines had begun buying them in huge quantities for their ground units. The next launch of a GPS satellite was scheduled for 2 August 1990.

So the elements that would have a bearing on CSAR in Desert Storm were in place. The Air Force, through a series of reorganizations had finally created the ARS, which had the CSAR mission, but it did not have the force structure to accomplish the combat recovery portion of the mission. Its primary recovery aircraft were antiquated and in the process of being replaced. The new aircraft were arriving slowly and were being parceled out to several units. The newly formed AFSOC had the force structure and a large part of the corporate combat recovery memory, but its units no longer had combat recovery in their DOC statements. The Navy had formed units tasked for combat rescue, but they were in the reserve component and needed presidential authority to be activated.

The Air Force—the service most likely to need CSAR—was relying on an untested concept to reliably and accurately locate its downed airmen, especially in high-threat areas where they were most likely to be shot down. And, for the most part, the

Air Force and Navy were equipping their aircrews with a single survival radio that was inferior to those used in the last days of the war in Southeast Asia.

In the summer of 1990, CSAR in toto was not in the best of shape. Regardless, it would soon be time to go to war.

Notes

1. Lt Col Joe E. Tyner, "AF Rescue and AFSOF: Overcoming Past Rivalries for Combat Rescue Partnership Tomorrow," National Defense Fellows Program Study (Maxwell AFB, AL: Headquarters US Air Force and Air University, undated), 8.
2. Capt Jerry Shipman, interview by the author, 13 June 2002.
3. Letter to author from Col Gary Weikel, USAF, retired, subject: Unofficial White Paper on the US Air Force Pave Low, 2000.
4. Shipman, interview.
5. Weikel, letter.
6. Shipman, interview.
7. Tyner, study, 12.
8. Col James H. Kyle, USAF, retired, *The Guts to Try* (New York: Orion Books, 1990), 24–27. Amplifying comments by Col Tom Samples, 24 July 2003.
9. *Aerospace Rescue and Recovery Service 1946–1981: An Illustrated Chronology*, pamphlet, undated. Amplifying comments by Col Tom Samples, 24 July 2003.
10. Tyner, study, 13; and comments to the author by Col Tom Beres, USAF retired, 15 October 2003.
11. Tyner, study, 14.
12. Kyle, *Guts to Try*, 224.
13. Ibid., 360.
14. Shipman, interview.
15. Shipman, interview; and Kyle, *Guts to Try*, 254.
16. Tom Clancy with Gen Carl Stiner, *Shadow Warriors* (New York: G. P. Putnam's Sons, 2002), 9.
17. Stovall, interview by the author, 3 September 2001.
18. Sid Balman, "Second: US Force Planned to Invade Tehran to Free 52," *Air Force Times*, 25 September 1989, 24.
19. Weikel, letter; and Stovall, interview.
20. Tyner, study, 18.
21. Shipman, interview.
22. Ibid.
23. Susan L. Marquis, *Unconventional Warfare* (Washington, DC: Brookings Institution Press, 1997), 77.
24. Maj Gen William J. Mall, "Commander Shares Insights," *Airlift Magazine*, Fall 1984, 1–3.
25. Shipman, interview.

26. Stovall, interview; and Tyner, study, 24.
27. Col Anthony Burshnick, interview by the author, 13 September 2002.
28. History of the Twenty-third Air Force pamphlet; and comments by Col Tom Beres, retired, 15 October 2003.
29. Weikel, letter. This issue came up numerous times and is, in fact, a simmering disagreement between the Army and Air Force special operations communities. Also, Lt Col Richard Comer, interview by the author; and e-mail from Gary Weikel to author, 5 September 2002.
30. Tyner, study, 28.
31. Ibid.
32. Ibid.
33. E-mail from Lt Col John Blumentritt to author, 28 August 2002.
34. Mr. Ted Lunger, interview by the author, 23 July 2002.
35. Ibid.
36. Ibid.
37. Tyner, study, 31.
38. US Special Operations Command (SOCOM) 10th Anniversary History, HQ USSOCOM/SOCS-HO, MacDill AFB, FL, November 1999.
39. Comer, interview.
40. Kyle, *Guts to Try*, 89.
41. Comer, interview.
42. Weikel, e-mail to author, 8 August 2002.
43. SOCOM History, November 1999, 17–20.
44. Stovall, interview.
45. Military Air Command (MAC) History CY 90, vol. 1, 10.
46. Lt Gen Buster C. Glosson, interview by the author, 25 September 2002.
47. MAC History CY 89, vol. 1, xxxv.
48. US Air Force Rescue Force Structure Plan (RFSP), Support Document I-79 to MAC History CY 89.
49. Tyner, study, 33.
50. Ibid., 35.
51. History of the Air Rescue Service, 1 January 1989–31 December 1990.
52. MAJ Mark Tucker, USA, interview by the author, 6 June 2002.
53. RFSP, B-2.
54. MAC History CY 89, vol. 1, 37.
55. CAPT Michael T. Fuqua, USN, "We Can Fix SAR in the Navy," *Proceedings*, September 1997, 57.
56. Shipman, interview.
57. Michael S. Breuninger, *United States Combat Aircrew Survival Equipment* (Atglen, PA: Schiffer Military/Aviation History, 1995), 160.
58. RFSP, 6-1.
59. Stovall, interview.
60. Michael R. Gordon and Gen Bernard E. Trainor, *The Generals' War: The Inside Story of the Conflict in the Gulf* (New York: Little, Brown and Co., 1994), 250.
61. Reported by several pilots who flew in Desert Storm.

62. International Civil Aeronautical Organization (ICAO) Circular 185, *Satellite-aided Search and Rescue—COSPAS-SARSAT System* (Montreal, Canada: ICAO, 1986), 17.

63. Summary of US Air Force Aircraft Losses in Southeast Asia, 12 January 1962–31 August 1973, K417.042-16.

64. Comer, interview. See also RFSP, 6-1. For an explanation of the GPS and its use in Desert Storm, see Michael R. Rip and James M. Hasik, *The Precision Revolution GPS and the Future of Aerial Warfare* (Annapolis, MD: Naval Institute Press, 2002).

Chapter 3

Desert Shield

We came to do a job, and it's a worthwhile job. So as far as I am concerned, we will all stay until that job gets done.

—Lt Gen Chuck Horner

The Iraqi invasion of Kuwait on 2 August 1990 came as a surprise to most Americans. It had been a pleasant summer and most thoughts were on vacation, baseball, and football—anything except war. The graphic news reports caused people to stop and wonder what this all meant. Would there be gas lines again? Would this mean war? Would this mean a military draft?

But to those individuals charged with protecting American lives and interests in that part of the world, it came as no surprise at all. In fact they had expected it for some time. The US Central Command, or CENTCOM, had been reporting that the possibility of such an Iraqi move was steadily increasing. The prior day GEN Norman Schwarzkopf, CENTCOM commander, had briefed Secretary of Defense Richard "Dick" Cheney that an attack against Kuwait was imminent.[1]

Our nation's response was swift in coming. Pres. George H. W. Bush immediately took economic and political steps to punish Iraq. In discussions with his senior advisors, he established clear political objectives:

1. Remove the Iraqis from Kuwait.
2. Eliminate production and storage of weapons of mass destruction.
3. End Iraq's capacity to threaten its neighbors over the next five to 10 years.
4. Ensure that the full conventional military capabilities of the United States would be used.

They also established a crucial limiting factor: the desire to hold American military and Iraqi civilian casualties to a minimum.[2] This last factor was critical because recent history had shown that the American public had little tolerance for the loss of American lives. This would put a premium on having a CSAR capability in-theater.

As the president was making these decisions, orders went out to military units across the land to prepare for deployment. Two aircraft carriers, the USS *Independence* and the USS *Dwight D. Eisenhower*, and their supporting task forces immediately sailed for the area.

Almost unnoticed in the rush of events was a report that GPS satellite PRN-021 had been successfully launched into orbit from Vandenberg AFB, California. Its orbit would specifically cover the Persian Gulf area.[3]

Due to growing concern that Iraq might take drastic steps, CENTCOM had recently held a war game at Hurlburt Field, Florida, to explore how such an event would unfold—and its repercussions. Called Internal Look, it presupposed such an Iraqi attack against Kuwait. This was a radical change for the command, because for years its fundamental war plans had assumed the main threat of war in the region would be an attack by Soviet forces through Iran to seize ports in the Gulf region.

Fortuitously, Internal Look forecast how such an Iraqi attack could occur and suggested that CENTCOM needed to make major changes to its primary war plan, OPLAN 1002-90. It would also require a complete recalculation of the force deployment plan called the time-phased force and deployment list, or "TPFDL," as the planners called it. Such changes took time. The staff at CENTCOM began the lengthy process of changing it. The process was expected to take about a year.[4]

An annex buried within the massive plan laid out broad planning for CSAR operations. Its author was Army LTC Pete Harvell. He was a career special operations officer and worked in the J-3 operations branch of CENTCOM, where he was the primary staff contact for CSAR. Using the changes generated by Internal Look, he updated the CSAR annex to the war plan.[5]

On 5 August Secretary of Defense Dick Cheney and Secretary of State James Baker traveled to Saudi Arabia to meet with King Fahd. They graphically laid out for him the danger that faced his nation. Two days later the government of Saudi Arabia formally asked for US military assistance.[6] General Schwarzkopf was directed to execute his war plan. The current TPFDL

Col George Gray was selected to be the commander of AFSOCCENT, the Air Force component of SOCCENT.

was activated to direct the scheduling and flow of forces to the area. It identified the units, personnel, and equipment that would deploy. Over the next several months, more than 500,000 troops, 1,800 aircraft, and millions of tons of supplies and heavy equipment would flow from around the world to the Gulf region.[7]

Units of all services were identified for deployment to CENTCOM. Air Force units deploying would be assigned to the CENTCOM air forces component, or CENTAF; Army units to the army component, ARCENT; Navy forces to the naval component, NAVCENT; and Marine forces to MARCENT.

Since the creation of SOCOM, the Central Command now also had a sub-unified command assigned from SOCOM. This was called the Special Operations Command Central or SOCCENT and was commanded by Army COL Jesse Johnson, a veteran of the failed Eagle Claw operation in Iran in 1980.[8]

Forces were designated to be assigned to SOCCENT for contingency operations from the special operations branches of the Army, Navy, and Air Force. The Air Force units were from the newly created AFSOC at Hurlburt. As the TPFDL began to run, orders went out to units across the land. Some of the first to receive orders were the units assigned to the AFSOC, primarily in the 1st SOW, also at Hurlburt. In fact, its commander, Col George Gray, was also designated to be the commander of AFSOCCENT, the Air Force component of SOCCENT. As such he would be Colonel Johnson's head airman and would also assume the operational duties of the joint special operations air component commander or JSOACC. Colonel Gray took with him Col Ben Orrell, his director of operations—the same Ben Orrell who had received the Air Force Cross for the Bengal 505 rescue in Laos in 1972. When it came to combat rescue, he was the expert.[9]

As per the war plan, personnel from the various units at Hurlburt were quickly recalled and ordered to prepare for deployment. Maintenance and support personnel worked feverishly to get the aircraft and equipment ready to go. Personnel processing and vaccination lines were set up to take care of the necessary last-minute details prior to leaving. Several squadrons from the 1st SOW were identified to go: the 8th SOS with its MC-130E Combat Talons, the 9th SOS with its HC-130P/N Combat Shadows, the 20th SOS with its MH-53J Pave Lows, the 55th SOS with its MH-60G Pave Hawks, and the 1720th STG that commanded combat control teams and pararescuemen.[10]

As the TPFDL ground on and identified the units to be alerted, one deficiency jumped out at the planners—the rescue units identified to support OPLAN 1002-90 were not available. They had either been deactivated or were not shown as combat-capable. In fact the Air Rescue Service was deep into its conversion from HH-3s to the vastly improved HH-60Gs coming off the production line. With personnel from several units training on the new aircraft, ARS did not have any deployable, combat-ready HH-60–equipped units to send. As one rescue pilot noted at that time, "The Gulf War could not have occurred at a worse time for the Air Rescue Service."[11]

At least one HH-3 unit was ready though. The 71st Air Rescue Squadron in Alaska was combat-ready and willing to go, but its commander, Lt Col Larry Helgeson, never learned why his unit was not called. He noted,

> The special operations forces units were more combat ready in that they were better funded with better equipment and were better capable of going in and doing it [combat recovery]. It [HH-3] was an older airplane. It did not have the capabilities of what the special operations forces aircraft had. . . . The HH-3 was a dated bird. It had its mission, but it was in its prime back in the '60s. It filled a gap and did provide a capability. But it was not the combat deep-penetration rescue vehicle. The crews trained in it so they would keep those skills alive. But the machine was no longer viable.[12]

Regardless of the state of readiness in the ARS, CENTCOM needed a force to do the combat rescue portion of CSAR. Fortunately, there were Air Force officers in both CENTCOM and SOCOM who knew that the combat rescue capability still resided in the AFSOC helicopter and HC-130 communities. At the time, Brig Gen Dale Stovall, who as a young captain had received the Air Force Cross for the dramatic Roger Locher rescue in North Vietnam in 1972, was serving as the vice commander of AFSOC. Since his days as a fired-up young HH-53 aircraft commander in Thailand, he had continued to mature into another noted giant in the combat rescue community. He recalled that "the physical equipment capability and the operational training and experience all resided within the MH-53 community."[13]

There was still some skepticism about the efficacy of CSAR at the highest levels of the Air Force. The future chief of staff, Gen Merrill McPeak, made the comment—apparently more than once—that in a high-threat war, combat rescue was perhaps too dangerous. He was worried that the rescue crews would take too many risks and become casualties themselves. "I don't want to trade three for one," he stated, apparently thinking back to his Vietnam days when he erroneously recalled that a helicopter crew generally consisted of a crew of three.[14]

Earnest discussions took place between the various headquarters, both in the United States and in the theater. General Stovall described how it worked:

> When you go into theater under the 1986 Congressional Act [Goldwater-Nichols], the theater commander owns all the assets. Period. He can do anything he wants with those assets. So they made the decision in the theater that they were going to use them [the AFSOC helicopters] for combat recovery . . . probably [General Horner] was involved in it or somebody else there who said, "Hey, we have to have SAR. We don't have SAR capability within the Air Force. We need to go to Special Ops Command and get them involved in it."[15]

Other agendas played into the assignment. At CENTCOM J-3, LTC Pete Harvell knew that Schwarzkopf harbored some hard feelings towards special operations troops and had no enthusiasm for moving them quickly into the theater. Well familiar with the war plan, he knew that significant special operations forces were not due in until 90 days after the initial deployment of troops. Harvell saw combat rescue as a way to get special operations forces in early and pushed for SOCCENT to get the tasking.[16]

This mission assignment to SOCCENT met some resistance in SOCOM. Some Army officers in particular were upset about using the MH-53s and MH-60s for combat recovery. Several lamented the fact that SOCOM had put a lot of money into upgrading and operating these helicopters to support their specific missions. All of that had paid off handsomely the previous year in Panama during Operation Just Cause. Now it appeared that a little "sleight of hand" was being used to grab them for the combat recovery mission at the expense of other classic special operations missions. General Stovall listened to the bickering and then responded, "Okay, you tell the President that you want to hold these back for special ops missions that we are currently not flying rather than going out and picking up somebody who has gotten shot down. You tell him."[17] That ended that discussion. Col George Gray at the 1st SOW remembered it another way. "We got the rescue mission by virtue of our being the only ones left out there that had the weapons system that was capable of doing it," he recalled.[18]

None of that really mattered to the men of the 20th SOS at Hurlburt. Their instructions, as relayed by their commander, Lt Col Rich Comer, were to get their aircraft ready to go and get on the big cargo planes for the ride over. Comer knew that combat recovery would be their priority mission. As an old rescue guy, it suited him fine. He knew that, initially at least, about the only significant US combat power in the region would be airpower. It would be responsible for providing the initial defense of Saudi Arabia. His MH-53s would be the initial combat recovery force.[19]

The forces for combat recovery would be under the command and control of SOCCENT. That was the command relationship that the combatant commander, General Schwarzkopf, wanted, and under the laws of the United States, it was his call to make.[20]

As the AFSOC forces began to flow into the theater, the AFSOC commander made a courtesy call to the ARS commander to inform him that Schwarzkopf had decided that the special operations units being attached to SOCCENT would be assigned the responsibility for combat rescue.[21] This did not sit well with many in the now-regenerating rescue community. Even though their HH-53s had been taken away and replaced with older HH-3s, they trained diligently for the mission and were enthusiastic to go. They knew that their old aircraft had limitations, but they still wanted to deploy and provide what services they could. After all, it was their job. Regardless of what the operational orders might say, THEY were the rescue guys. If rescue were going to be needed, they wanted to be sent. A few HH-60s had been delivered, and crews were being trained. Additionally, some HH-60s had been delivered to Air National Guard units. Already, members of those units were volunteering to go. Lt Col Mark Tucker, now on the Air Staff, recalled that,

> I was in SAF/AQL, classified acquisition. . . . I was appalled when I heard that we were not sending any rescue units over. We could have taken crews and aircraft from several units and sent them over to form a provisional unit. ARS [commander John Woodruff] said, "No, we are not going to do that." We could have taken the lead crews . . . in order to form a combat-capable unit. . . . I will never forget being absolutely appalled at the decision not to take those crews and aircraft to marry them together to form a combat-capable unit. Following 10 plus years after Desert One, having watched that, I just couldn't conceive that rescue [ARS] would be willing again not to ante up the forces to go where the fight was.[22]

Still, it wasn't that simple. SOCCENT was glad to take the mission. Like Comer, their commander Colonel Johnson knew that the war would include some form of air campaign and that projections showed that many aircraft would be downed. He intuitively understood that his troops were the right asset for rescue. There was a larger issue, too. He knew that Schwarzkopf had an inherent distrust of the special operations forces and was hesitant to give them any significant missions. Combat rescue was their way of getting into the theater early, even though it was not one of their primary missions.

General Schwarzkopf acquiesced to their deployment, but he forbade Johnson from launching any cross-border operations without his express approval because he feared that such actions could trigger a war before he was ready to fight it.[23]

Receiving the tasking, Colonel Johnson saluted smartly and made combat rescue his command's top priority.[24] In fact, the first assigned mission to the Army Special Forces teams and Navy SEAL teams was to be ready to conduct recovery operations.[25]

On 9 August AFSOC was directed to begin sending advance personnel to CENTCOM and SOCOM to do the endless coordination necessary for deployment. That evening, Lt Col J. V. O. Weaver, Lt Col Randy Durham, Maj Robert Stewart, and Capt Randy O'Boyle—all from the 1st SOW—left for MacDill AFB, Florida, to assume positions on the CENTCOM staff. Within a few days they had all departed for Saudi Arabia as part of the initial contingent. The first was a 40-man team from all three components. Arriving in Saudi Arabia, each group went its own way and began looking for a place to bed down its following units.

Colonel Weaver coordinated to have the arriving AFSOC units placed at Dhahran International Airport, Saudi Arabia. Unfortunately, that word did not get back to Hurlburt, and the first contingent of four MH-53s, crews, and support personnel left on two C-141s and a C-5 headed for Riyadh. Upon landing there in one of the C-141s, the team leader discovered his mistake and diverted the other two aircraft to Dhahran. After they landed and unloaded, he talked one of the C-141 crews into coming to Riyadh to pick up the rest of his team and deliver them to the correct location.

The CENTCOM staff also began to deploy from its peacetime headquarters in Tampa, Florida, to Dhahran. One of the first to go was LTC Pete Harvell along with other J-3 personnel aboard an early C-141. Upon arrival they immediately began to set up the huge command center necessary for combat operations. Harvell also began setting up the necessary CENTCOM structure so that arriving SOCCENT-assigned units could immediately assume rescue alert.[26]

Working feverishly in temperatures as high as 125°F, 20th SOS maintenance crews used borrowed tools and cranes to get the MH-53s ready to fly by 16 August. Within days, two aircraft a day were flying support for SOCCENT Army and Navy teams along the Kuwaiti border. Additionally, one aircraft was constantly on combat recovery alert, because intelligence indicated that, at any time, the Iraqi forces in Kuwait could attack across the Saudi border. In that case the MH-53s would need to fly up immediately and extract the special forces teams along the border. The helicopters were also available for any in-country SAR work. This was a major consideration because the number of coalition aircraft in the region was increasing daily, and more and more aircraft were flying local training sorties. Located not far from a Saudi helicopter unit, Comer had some of his men coordinate with the Saudis on local and regional recovery procedures.[27]

F-15s were now flying defensive patrols overhead daily to guard against surprise Iraqi Air Force attacks. The MH-53 crews were ready to mount necessary recovery efforts, even though they had not yet received machine guns or ammunition for the aircraft.[28] Four more MH-53s were in place by early September, and by midmonth the 20th SOS had it full complement of tools and spare parts.

In mid-August CENTAF activated its Tactical Air Control Center (TACC) in Riyadh. It rapidly began to fill with personnel who were initially focused on developing an immediate plan to deter and, if necessary, to stop an invasion of Saudi Arabia by the Iraqi military. They had to set up the TACC outside of the Royal Saudi Air Force (RSAF) headquarters in an inflated module that was immediately labeled "the Bubble."[29]

The commander of CENTAF, Lt Gen Charles Horner, at the direction of General Schwarzkopf, deployed to Riyadh to act as the "front man" for CENTCOM as the entire process kicked into high gear. He monitored the initial efforts to develop a defensive force and capability, but his mind was on offensive operations.

At Schwarzkopf's request an initial strategic air campaign plan had been developed by Col John Warden, the head of the "Checkmate" division on the Air Staff in Washington. Warden briefed the concept to Schwarzkopf and chairman of the Joint Chiefs of Staff, GEN Colin Powell, and was then sent to Saudi Arabia to brief General Horner. Horner accepted the briefing as the basis for an offensive campaign plan; but to turn it into an executable plan, he needed a planning staff in Riyadh. He grabbed Brig Gen Buster Glosson, a career fighter pilot, to head that effort.

Glosson had been assigned as the vice commander of a joint task force in the Gulf and jumped at the chance to join Horner and the growing CENTAF team. Glosson was a bit of an enigma. Personable and slow talking, he had spent time working in legislative liaison. The gentle demeanor that he had developed there belied the driven, hard-charging fighter-pilot attitude within. The responsibilities that he was accepting would require the use of both skill sets as he immediately began to build a small staff called the Special Planning Group. The group met behind closed doors, and the workers immediately dubbed it the "Black Hole."[30]

Lt Col Randy Durham and Maj Robert Stewart deployed from MacDill to serve as the SOCCENT liaison element in the TACC. Capt Randy O'Boyle was detailed to the CENTAF staff to act as the special operations liaison there. O'Boyle was a young MH-53 pilot and had grown up in the 20th SOS. The big Irishman was never at a loss for words on any subject. Some interpreted this as bombast, but those who spent time around him soon came to realize that he knew his stuff. He was intense in all he did—work, athletics, flying, partying—it was rumored that he considered any party a failure if the police did not arrive at least twice. He was, as another unit pilot said, "a piece of work," and he was ready to go to war.

Before leaving Hurlburt, O'Boyle was grabbed by Colonel Comer. He knew that Randy would somehow end up in the thick of things and made him promise that he would not commit the 20th SOS to anything until he discussed it with Comer. O'Boyle saluted smartly and headed out.

Between 20 and 22 August the 9th SOS arrived with four HC-130 aircraft. They settled at King Fahd International Airport (KFIA). Nine days later, Col George Gray, commander of the 1st SOW, arrived to assume his duties as AFSOCCENT and JSOACC.[31]

George Gray was another classic individual. Stout with a shock of red hair, he initially started out as a cargo pilot and transitioned to special operations assignments back in the '80s. He loved his job as commander of the 1st SOW and flew in all of the assigned aircraft as much as he could. He took a stern line with his troops and pushed them hard, but he also loved them and gave them the best care he could. Their fights were his fights. He knew that his unit would soon face hard combat. That was as it should be. But he did not intend to lose any of them—at least not needlessly.

Prior to leaving MacDill, O'Boyle had reviewed the command CSAR guidance currently in effect for CENTCOM. Arriving in the TACC, he set up a desk and, not having anything else to do, started working on a detailed combat recovery plan for the theater. He put up a map and began drawing lines into Iraq. His map caught the attention of one of the planners in the planning cell, Lt Col Steve "Foose" Wilson. Another career fighter pilot, Fuss was assigned to the fighter plans shop on the Air Staff in the Pentagon. When significant air units started heading for Saudi Arabia, he was dispatched to act as an Air Staff liaison to the CENTAF staff, where Glosson grabbed him and threw him into the Black Hole.

What went on in the Black Hole was a big secret because any planning for offensive action against Iraq was forbidden at this time. That was why the lines on O'Boyle's map drew Wilson's attention. Why, Foose asked, was O'Boyle doing what he was doing? O'Boyle flashed his best Irish smile and replied that he figured that they would eventually need an overall CSAR plan, and since he had been the plans officer in the 1st SOW at Hurlburt and knew a few things about CSAR, he would start the process.

Wilson then asked Randy increasingly detailed questions about the capabilities of various aircraft and how they would conduct combat rescue missions. Randy continued to explain. Finally, Wilson said, "Follow me," and took the young captain into a classified room, where he introduced him to Lt Col Dave Deptula. Deptula had come over with John Warden and had

DESERT SHIELD

Photo courtesy of Randy O'Boyle

Capt Randy O'Boyle and Brig Gen Buster Glosson

also been hijacked by Glosson for his team. He asked more questions. Sensing his moment, O'Boyle started talking. Deptula liked what he was hearing and said to O'Boyle, "We need you," and took him in to see Glosson. The process was repeated, and Glosson told O'Boyle that he was now part of the planning team in the Black Hole. O'Boyle was briefed into the detailed Top Secret planning to design an air campaign plan against Iraq and directed to step up his rescue planning and fit it into the larger effort.[32] In short order Glosson came to consider O'Boyle his "right hand man or even alter ego" for rescue.[33]

Captain O'Boyle was not working in a vacuum. CENTCOM had addressed CSAR in its theater war plan. The overall war plan for CENTCOM was USCINCCENT OPLAN 1002-90. It had been recently reviewed and updated by Colonel Harvell and the troops in J-3, and was dated 18 July 1990. ANNEX C specified basic policies and procedures for participation of US military forces in SAR activities within the CENTCOM area of responsibility (AOR). It also noted that no SAR forces were apportioned to CENTCOM for the plan.

It presented several key definitions:

1. Search and Rescue (SAR)—Use of aircraft, surface craft, submarines, personnel, and equipment to locate and recover personnel in distress on land or at sea.
2. Combat Search and Rescue (CSAR)—A specialized SAR task performed by rescue-capable forces to effect recovery of distressed personnel from hostile territory during contingency operations or wartime.
3. Joint Rescue Coordination Center (JRCC)—A primary SAR facility suitably staffed by supervisory personnel of two or more services and equipped for coordinating SAR operations.
4. Rescue Coordination Center (RCC)—A primary SAR facility suitably staffed by supervisory personnel and equipped for coordinating and controlling SAR operations at component level or higher.
5. SAR Coordinator (SC)—The designated SAR representative of the area commander, with overall responsibility and authority for operation of the JRCC and for joint SAR operations within the assigned geographical area. (For this OPLAN, the SAR coordinator was General Horner).[34]

The concept of operations stated that General Schwarzkopf as the area commander had primary responsibility and authority for SAR in the CENTCOM AOR and would ensure the development of plans and procedures for the effective employment of all available SAR resources.

The various component commanders—COMUSARCENT, COMUSCENTAF, COMUSMARCENT, COMUSNAVCENT, and COMSOCCENT—would provide organic SAR forces as appropriate and would provide supervisory liaison personnel to the JRCC to effect required coordination of the SAR effort for all US forces. All component commanders could use organic resources for initial efforts to recover personnel.

Additionally, component/supporting commanders would

1. Be prepared to support and/or conduct SAR operations when feasible or as directed by the SC.

2. Establish an RCC to facilitate the execution of SAR operations.

3. Provide the JRCC with SAR liaison officers (SARLO) as appropriate.

As the SC, Horner had different tasks. Specifically, he could

1. Direct components to conduct SAR operations on a mission priority basis. He also had authority to use assets from any component for SAR operations, assuring that assets required for SAR were not diverted from component commander missions with a higher USCINCCENT-established priority.

2. Maintain the JRCC at the TACC and, through MAC, provide the nucleus of SAR controllers. Provide the senior ranking officer facilities and communications for the US-CENTCOM.

3. Establish an RCC, as appropriate.

4. Coordinate with the JRCC for SAR requirements and advise the JRCC of any unilateral actions taken by forces in the prosecution of SAR missions.

Under his direction, the JRCC had the authority to

1. Control and coordinate SAR operations within the AOR. Operational control was exercised by the SAR coordinator through the JRCC and the component RCC/SAR controllers.

2. Task supporting/component commanders with SAR missions, as directed by the SAR coordinator.

3. Maintain close liaison and coordination with forces of other nations, international agencies, and appropriate DOD or CENTCOM representatives on SAR activities in the AOR. Where international SAR forces were available and capable of lending assistance in SAR operations, coordination would be effected by the JRCC to incorporate/utilize the assets made available by these forces.

4. Develop SAR plans.[35]

Cascading down from this, COMUSCENTAF had published an OPLAN 1002-90 and OPLAN 1021-90 BASE CASE that amplified and clarified more specific procedures.[36]

Captain O'Boyle was also not the only AFSOC troop working on CSAR. In August Maj Ken Black of the 8th SOS worked with several CSAR specialists to develop an initial plan directing CSAR operations during Desert Shield. It was, in part, consolidated into the long-term CSAR planning that Captain O'Boyle was doing. That plan divided the AOR into four regions: north, west, central, and east.

At the same time, SMSgt Dan Hodler and TSgt Scott Morrison from the 1720 STG, and TSgt Ken Matney, an Air Force survival instructor from the Survival School at Fairchild AFB, Washington, began physically setting up the JRCC within the TACC. They also began another effort to build a theater CSAR plan.[37]

As these efforts were progressing, CENTAF sent a request to the ARS for augmentation to fill the JRCC. In early September, these men were joined by a team of nine officers and enlisted personnel from the AFRCC at Scott AFB. The leader was Lt Col Joe Hampton, a navigator and career rescueman, currently assigned as the chief of rescue coordination operations. They were joined by another group from Ninth Air Force out of Shaw AFB, South Carolina. It was headed by Maj Joe Stillwell.

As the senior officer, Hampton would be the overall director. He had to meld the two groups into a team. He did not anticipate that this would be a problem since most of the personnel had long rescue backgrounds. Yet none had any actual CSAR experience. Hampton himself had entered the Air Force in the early '70s but missed any service in Southeast Asia.

Hampton was a little worried because there was so much to do. He took steps to get all of his people the necessary security clearances, requested the communications and phone lines that they needed, and instituted an internal CSAR training program. Overall, Hampton was comfortable with the setup. He had worked with the men and women from Shaw on several exercises focused on the Middle East and was very familiar with the theater.[38]

Hampton assumed that prior to actual combat operations, his team would have opportunities to take part in several war games and local exercises. Given the joint and ultimately combined nature of their work, Hampton also went to work to get representatives from each of the services to work directly in the JRCC. He also asked for a Saudi and a British representative. Over the next months all of his requests were met.

After conducting a physical inventory of the JRCC set up by the AFSOC troops, Hampton was not satisfied. The center had only one phone, and it could not be used for classified calls. They had to go to another section to make radio calls. Their dedicated equipment pallet had not arrived.

Shortly after arriving, Hampton met with Horner. The general made it clear that Hampton worked for him and that he personally put a lot of priority on CSAR. He also told Hampton that anytime he needed anything, he was to go directly to the colonel running that shift. If that did not solve the problem Hampton was to come to him. Hampton saluted smartly and got back to building his operation.[39]

At about the same time Hampton and his men were unpacking their bags, the 55th SOS began arriving with their eight MH-60 Pave Hawks. Within a week, the crews were taking orientation flights. A few days later they were working with ground teams from the Navy and Army. The SOCCENT fleet of helicopters continued to expand and settle into their missions.[40] They were steadily augmented, not only with PJs arriving from the 1723d Special Tactics Squadron, but also with troops from units all over the globe.

In early September, as Hampton and his team were getting settled in, he called for a meeting in Riyadh of all units and personnel who would be involved in CSAR. O'Boyle called Comer from the 20th SOS and told him about it. Comer attended as did several other commanders and planning representatives. He was pushing his squadron hard to be ready and needed to know what their particular combat rescue responsibilities were going to be. He knew that the JRCC had been formed and needed details so he could plan, allocate assets, and assign troops. The meeting did not answer his questions. He recalled,

> I was talking to Joe Hampton, the guy holding the meeting. What he initially had was Saudi Arabia cut up into sectors, sections, and circles. Circles were based on aircraft ranges from bases, where the Saudis were, and Air Force and Army helicopters. It was, "You guys are going

to do this and you guys are going to do this, and you guys are going to handle it if it is here," and it was basically a peacetime SAR [plan]. . . . There were no capabilities-based divisions to that. It was just, "If you've got a helicopter, will it go this far?" THERE WAS NOTHING COMBAT ABOUT IT. It was all peacetime. I said, "Well, this is all well and good. Where is our combat plan?" I think he said, "We don't have anything, we aren't ready to do that yet."[41]

Comer left the meeting very frustrated. He did not have the guidance he needed and would have to improvise. There was still so much to do.

Joe Hampton and his troops begged and scrounged for material. They lacked everything. In fact, about all that they had initially were a few laptop computers for administrative items. While building his operations center, he discovered that the planning efforts of O'Boyle and the others were well advanced. Working with them, he completed a theater CSAR plan by late October. Additionally, SOCCENT had formalized its plans.

In November the JRCC published the detailed instructions for CSAR in the AOR titled, *Operation Desert Shield Combat Search and Rescue Plan*, dated 1 November 1990. It was signed by Horner and his Saudi counterpart, Lt Gen Ahmed Behery. Its purpose was to "provide a combined Personnel Recovery (PR)/Combat Search and Rescue (CSAR) Plan which will integrate into a cooperative network available to Saudi Arabian, US, and other capable and interested parties. The available resources will be coordinated by a single combined agency in order to afford greater protection of life and ensure efficiency and economy."[42]

In this plan the JRCC was designated as the combined agency for SAR operations in Desert Shield. It listed, in detail, its own responsibilities and those of its tasked units. The JRCC would consider search missions in which the location and condition of the objective(s) [survivor(s)] were unknown, or rescue missions in support of survivors or other persons in distress whose locations and nature were known. Additionally, they were prepared to coordinate rescues throughout the AOR, which included Iraq, Kuwait, the Persian Gulf and its adjacent countries, the Red Sea, the Gulf of Aden, and a slice of the Arabian Sea down toward Diego Garcia.

Each service component was then tasked by the higher headquarters' OPLANs to establish its own RCC and be prepared to

DESERT SHIELD

handle intracomponent rescue missions. Those missions beyond any component's abilities would be referred to the JRCC. Each component was also expected to have organic forces available for rescue missions at the JRCC's request, yet each would maintain operational control for launch of its own forces. Hampton explained that,

> They [each component] had launch authority. They had command and control of each asset for the mission. . . . As far as the tactics, the launching, what crews to use, how to use them—that was up to the command who owned and trained them like the [SOF] forces or the Marines if we tasked them or the Army, if we tasked them for this. It may have been different than what was in Vietnam, but at the same time, the environment, the way things were operating, rather than try to build up an organization saying "we own this stuff," well, we didn't have enough assets to do that. And it would take away from [their] operations.[43]

SOCCENT received the same tasking. But it would also be the primary combat recovery force theaterwide since its specially modified MH-53s and MH-60s were widely recognized as the best asset for recovery in enemy territory.[44] SOCCENT would maintain operational control (OPCON) over the launch of its assigned helicopters.

In coordinating this plan, Colonel Harvell fought hard to ensure that SOCCENT maintained operational control of the helicopters. He harbored an inherent fear that the Air Force operations officers in the TACC did not have a realistic feel for what the special operations forces could and could not do.[45]

Taskings were also included for each of the participating coalition forces. The Air Force RCC was, in fact, the JRCC. The Army RCC was located with ARCENT headquarters. The Marine RCC was located in the Marine Tactical Air Control Center (MTACC) at Shaikh Isa, Bahrain. Since the Marines as a service did not "officially" do CSAR, they had no specific RCC. In fact, they had TRAP (tactical recovery of aircraft and personnel) teams designated and some H-46 and UH-1 helicopters. They were kept on call by the MTACC.[46]

The Navy had its NRCC (naval rescue coordination center) with NAVCENT aboard the USS *Blue Ridge* in the Persian Gulf. Under the NRCC, Combined Task Force (CTF) 151 served as the regional SAR controller for Battle Force Zulu in the Persian Gulf and was responsible for planning CSAR operations over water. It had the authority to directly coordinate with SOCCENT. CTF 155 operated as Battle Force Yankee in the Red Sea. There, a SARC (search and rescue coordinator) was located aboard the aircraft carrier USS *John F. Kennedy* and coordinated the rescue actions of that area.[47]

The Saudi military had its RCC equivalent in its air combat operation center (ACOC). The Royal Saudi Air Force had helicopter squadrons throughout the country and was primarily responsible for in-country rescue.

The British had a rescue center set up with its Nimrod detachment at Seeb, Oman. Coordination took place through the Maritime cell in the CENTCOM combined headquarters in Riyadh.

The French Air Force kept a liaison officer in the TACC who could immediately contact his forces. They had a detachment of Puma helicopters at Al Ahsa Air Base in Saudi Arabia.

Every service component except the Air Force had helicopter assets in-theater that could be used for recoveries. Since the JRCC did not have operational control of any of the component assets, neither Hampton nor his controllers could order them to launch. They could ask, but they could not order.

Colonel Gray became a little concerned during the detailed CSAR planning. He was being told that SOCCENT was responsible for CSAR. That was impossible. CSAR is a process that includes locating survivors, marshaling forces, and then going in and making a recovery. His forces could not locate survivors. They did not have control of the assets to do that. Nor did they have the authority to direct other forces for rescue operations. All of that had to be done by the JRCC. He made it clear that the helicopters and crew under SOCCENT were CR (combat rescue) assets.

While the DALS equipment was on board some helicopters, it did not work well at low altitude, which is where the helicopters had to operate in high-threat areas. CSAR history clearly showed that helicopters were at high risk of being shot down in high-threat areas if they attempted to remain there for any amount of time. Neither he nor his commander, Colonel Johnson, was going to allow his helicopters to be sent into high-threat areas to run search patterns. Colonel Gray said that, "If you run across a platoon of deployed guys, they can shoot your butt right out of the sky and God only knows we couldn't track divisions, much less platoons. A squad could put maybe a couple of rounds in you. Hell, a platoon, 35 or 40 guys with AK-47s, would just eat you up."[48]

He explained that before committing one of his helicopters for a combat recovery, he needed to have an accurate location for the survivor(s). It was a critical factor in deciding whether or not to commit an aircraft and its crew.[49] Additionally, he told Horner that if someone were down in a high-threat area, he wanted the downed crewmembers to evade/hide until night when the helicopters would have a reasonable chance of recovery and survival.[50]

He and Colonel Johnson laid out three criteria for committing a helicopter for a combat rescue:

1. The location of the survivor(s) is known.
2. Evidence of aircrew survival:
    a. Visual parachute sighting, and/or
    b. Voice transmission from the crewmember and authentication.[51]
3. A favorable enemy threat analysis.[52]

Capt Tim Minish, another Pave Low pilot, had been assigned to work with the intelligence section in the TACC. He joined with O'Boyle to write the SOCCENT annexes to the CSAR plans. They designed a series of "spider routes" or predesignated routings for use by the SOCCENT helicopters for flying into Iraq and Kuwait. These routes and the points that defined them could be preloaded into the navigation computers of the helicopters. This could save precious minutes when helicopters had to be scrambled to recover downed crews.[53]

Capt Corby Martin, an MH-53 standardization and evaluation pilot from the 20th SOS, joined O'Boyle and Minish in their efforts. Together, they continued to refine helicopter operations. The Iraqi-Kuwaiti portion of the AOR was split into four sections: west, central, east, and north. SOCCENT helicopters would cover the west, central, and east. In the west sector, helicopters would be kept at Al-Jouf and forward at ArAr. In the center section they would be placed at King Khalid Military City (KKMC) and Rafha. In the east, they would reside at Ras Al Mishab on the coast.[54] CENTCOM had assurances that US aircrews would be safe in Iran, but nobody put much stock in that. Jordan was considered hostile, with intelligence indicating that downed crews there would either be turned over to the Iraqis or possibly be at physical risk because of the hostilities of the Palestinians.[55]

The north sector was almost 900 miles away. They realized that to cover that section, SOCCENT would need augmentation, preferably by rescue elements in Turkey. That was a problem that they could not solve. Captain O'Boyle did express his concerns to Glosson, who listened but had neither the authority to commit forces there nor the forces available to commit.[56]

Meanwhile, the helicopter crews were feverishly training for their missions. This included extensive night operations to include refueling as low as 500 feet above the ground. This gave the crews ample opportunity to get comfortable flying with their terrain-following radar and night-vision devices across the broad expanses of Saudi Arabia.[57]

Continuing to build up the JRCC, Colonel Hampton was concerned about their ability to locate downed airmen. He was not comfortable with the PRC-90 radios issued to the flyers and addressed his concerns to one of the colonels on the TACC staff. Hampton told him,

> If we are going to locate these people, we have to do certain things. The PRC-90 sucks. Get rid of it. I want PRC-112s for all the guys going in country. And I want to get some sort of [RC-]135 or a [British] Nimrod, or maybe an A-10 or somebody who is going to have the interrogator [DALS] on board that aircraft. I want him up there, flying around, on the border if necessary so that when a guy goes down, we've got some kind of location. [I was told] we can't afford it. . . . So there was a problem there of being able to locate a guy. So we would rely on Rivet Joint, AWACS, wingman reports, [or] intelligence gathering from Iraqi communications.[58]

As the air campaign plan developed, the ever hard-charging O'Boyle proposed to Colonel Gray that they set up "laagers," or protected camps, at various locations in Iraq. These defended positions would be staging points for helicopters and crews sitting on alert for recovery missions. They would reduce response time for what were expected to be heavy losses. The idea was discussed with LTC Pete Harvell on the CENTCOM staff. He liked the idea, but in considering its ramifications, he realized that to do so on any kind of permanent basis would be extremely costly in terms of personnel, logistics, and wear and tear on the helicopters just ferrying everybody back and forth. Colonel Gray considered it too, but eventually turned it down, deciding that it was far too risky and put the crews at too high a risk. Yet he was pleased at the aggressiveness of his young airmen.[59]

Navy Reserve

In November President Bush directed his commanders to begin making plans for offensive operations. This would necessitate an increase of US forces in the region. Using the authority granted them by President Bush in his presidential selected reserve call-up, the Navy activated its two reserve rescue units, HCS-4 (Red Wolves) and HCS-5 (Firehawks). The reservists reported for duty at their respective bases 30 November. In early December, 97 officers and enlisted troops departed with four of their SH-60Hs for Tabuk, Saudi Arabia, where they would operate as a consolidated unit, HCS-4/5 Spikes (specially prepared individuals for key events), and collocate with an RSAF rescue unit.

In place and set up by 14 December, the unit was under the operational control of CTF 155, the Red Sea battle group. Their first assigned mission was to sit alert with Navy SEALs (Sea, Air, Land—US Navy special forces) from SEAL Team Two as a maritime interdiction force. Several exercises were held, but no actual missions were flown.

On 16 January the unit was assigned to tactical control (TACON) of SOCCENT and redeployed to Al-Jouf. Immediately, two aircraft were dispatched to the forward field at ArAr to sit CSAR alert. Colonel Comer from the 20th SOS started working with them and was immediately impressed with their enthusiasm and professional approach to their duties. Additionally, the unit was augmented with a SEAL detachment equipped with fast attack vehicles (FAV) for possible land operations or to serve as augmentees on the helicopters. The Spikes maintained this status with aircraft on alert 24 hours a day for the next 51 days.[60]

Air Force Reserve

Under the same presidential recall authorization, the Air Force in December directed the call-up and mobilization of the

HH-3–equipped 71st SOS located at Davis-Monthan AFB, Arizona. This was a first-class unit, which had traditionally trained hard to include becoming proficient flying with night-vision goggles (NVG). Its 150 personnel reported for duty on 21 December and were immediately processed onto active status. While waiting for deployment, the unit HH-3s received several equipment modifications that greatly upgraded their capabilities. Consequently the aircraft were changed from an "H" designation to an "MH" designation. Under the command of Lt Col Robert Stenevik, unit personnel began deploying to Saudi Arabia on 10 January. Four days later the unit was in place with five MH-3 aircraft ready for tasking under the operational control of SOCCENT. For the next 51 days, unit aircraft and crews would be on alert for combat recovery tasking, either out of their bed-down base at King Fahd International Airport or at the recovered Kuwaiti barge *Sawahil* docked at Ras Al Mishab. They were also available for any other tasked missions. Given their ability to land on water, they were especially sought out for missions in support of SEAL teams.[61]

Army

In early September the 3d Battalion of the 160th (3d/160th) Special Operations Aviation Regiment (SOAR) deployed its MH-60s to KKMC. Attached to and under the operational control of SOCCENT, they were tasked by Col George Gray, the JSOACC, as were the assets of AFSOC as they arrived. Experts at nighttime small-team insertions and extractions, they could also be used on combat recovery. This was not normally a mission assigned to them, but working with men from the 20th SOS, they began to train to the mission. They were reinforced with a security and medical detachment from the 2d Battalion of the 5th Special Forces.

During Desert Shield the unit trained for combat rescue and developed the capability, especially at night, using NVG. Additionally, during the initial phase of Desert Storm, they maintained 50 percent of their aircraft on combat rescue alert and were responsible for the central sector.[62]

Global Positioning Systems

In late November the US Space Command (USSPACECOM) launched another GPS satellite. This was the 23d launch in the program. Sixteen of the transmitters were now in orbit and capable of providing guidance in the Gulf region. This would ensure 22 hours of coverage a day. Unfortunately, three of the satellites began to develop mechanical problems. Working continuously, the men and women of the 2d Satellite Control Squadron, located at Colorado Springs, Colorado, kept all 16 of the GPS transmitters online and providing accurate navigational guidance.[63]

In the United States, crash orders had gone out for small lightweight GPS receivers (SLGR). More than 10,000 were delivered before Desert Storm. Almost half were shipped to the desert.[64] There the fielded forces of the coalition quickly became adept at using them. Navigational kits were also shared with coalition forces, who immediately realized how the devices made accurate navigation over the trackless expanses of the desert much easier.

The coalition aircraft had other navigational systems available, such as the long-range aid to navigation (LORAN) system, inertial navigation systems (INS), or Doppler as their primary navigational instruments. They were not nearly as accurate as GPS and/or had significant drift rates. Those using the GPS noticed that the maps issued to them were based on old surveys and had errors of as much as three miles.

As more and more units, ships, and aircraft were equipped with these advanced devices, commanders at all levels began to realize that GPS gave them unparalleled ability to navigate rapidly and accurately throughout the theater. But to be properly used, GPS had to be set to a common navigational reference or datum. In reviewing this, USSPACECOM directed that all units should use the World Geodetic Reference System 1984 (WGS84) as the common datum. This would ensure that two GPS units at the same location would read the same coordinates. This would not necessarily be the same as coordinates read with LORAN, INS, or other navigation systems.[65]

The Campaign Plan

By December Glosson and his team in the Black Hole had developed the overall air campaign plan. It would consist of four phases:

I. Strategic Air Campaign. This phase was designed to gain control of the airspace and attack Iraqi leadership; nuclear, biological, and chemical weapons; Scud missiles; and electric and oil infrastructure. By seizing control of the theater airspace, aerial forces could then prosecute the rest of the campaign. This would involve attacking the Iraqi Air Force, both in the air and on the ground, and destroying its command and control facilities and radars.

II. Air Superiority over the KTO (Kuwaiti Theater of Operations). This was, in fact, redundant with Phase I but included at the specific insistence of General Schwarzkopf.

III. Preparation of the Battlefield. This called for direct attacks on Iraqi fielded forces such as tanks, vehicles, and artillery. It also meant degrading the fighting morale of the Iraqi troops. The specific goal was to destroy 50 percent of Iraqi armor and artillery before the initiation of a coalition ground offensive.

IV. The Ground War. In this phase, the focus would be on providing direct support to the massive coalition ground units as they attacked and destroyed the Iraqi army—especially the Republican Guard units in and near Kuwait.[66]

Implicit in the planning was the expectation that SOCCENT helicopters would be available, even eager to go in and rescue the downed aircrews.[67] CENTAF had its CSAR plan in place. The necessary force was available and the troops were honing their skills.

CSAR Exercises

Over the next several months, the JRCC and SOCCENT conducted a total of six CSAR exercises, or CSAREXes. The first was in September and focused on command and control issues. Subsequent exercises increased in sophistication and included survivors placed in the desert and in the Gulf. Elements from all services and allies were involved. They worked diligently to iron out command and control issues and allow the various components to work together.⁶⁸

Different scenarios were used. Some involved single-seat aircraft; others used larger crews replicating the loss of something like a C-130. Exercise Desert Force in December focused on command and control. Exercise Search and Rescue in January exercised all units and control elements.⁶⁹

The exercises took place in the desert and the Persian Gulf and involved all allies. A-10s worked at leading in rescue helicopters. Navy SEALs practiced stealth operations. Air Force PJs and Army Special Forces troops practiced overland recovery techniques. In this scenario, an MH-53 would drop off a ground element at an offset delivery point. The team would then infiltrate, make the recovery, and exfiltrate to another point for recovery by the MH-53s. This technique could be used if an aircrew were down in an extremely dangerous area where coalition aircraft could not survive at low level. Most of these operations were conducted at night when the lumbering helicopters had the best chance of success.⁷⁰

Rich Comer thought the exercises were very useful, recalling:

> We had one full scenario CSAR with the A-10s. . . . It was daytime. . . . We expected to do CSAR at night. We wanted to wait until darkness if possible. We felt that the air threat was such that we pretty much had to for survivability purposes. . . . And those heat-seeking missiles were out there. If they saw us in the daylight, we were toast. We had IRCM [infra-red countermeasures] and all of that but we did not feel that we could go cruising around in the open desert with no cover. . . . I wanted to use MH-60s in the day. The MH-53 loses its advantage in the day, too big a cross section. The MH-60s were smaller. The 55th SOS and the Navy reserve unit had them. They [HCS-4/5] were an impressive bunch of guys. They really wanted to go. They were willing. But they were ill equipped, no FLIR, no night capability.⁷¹

Each exercise was digested and critiqued. Everybody was sensitive to the possibility of coalition troops being taken as POWs.

The president of Iraq, Saddam Hussein, sensed this too. In threatening the "mother of all battles" if we attacked, he said to the US ambassador, "Yours is a society which cannot accept 10,000 deaths in one battle."⁷²

The fear of significant American casualties was overriding and drove the planners and commanders in Riyadh. Early air campaign planning suggested coalition losses as high as 300 aircraft. The planning for the first night suggested that 10 percent of the aircraft sent would be lost. The wing commander of the F-111Fs expected to lose more because of the type of targets that his crews were being sent against.[73] General Glosson estimated that they should not lose more than 50 fixed-wing aircraft. If the losses were heavier, he felt that it would be the result of their failure, and in that case, President Bush would have been more than justified in firing all of them.[74] General Horner himself predicted the loss of 42 USAF aircraft alone.[75]

The planners also noted that when the Iraqis invaded Kuwait, the Kuwaiti air defenses—consisting of modern I-Hawk missile batteries and massed antiaircraft artillery (AAA)—had downed 39 Iraqi aircraft in that short but intense battle.[76] The principal author of the initial air campaign plan for the war, Colonel Warden, estimated that 10 to 15 aircraft would be lost the first day and then no more than 40 overall.[77]

Reasonable people could logically or rationally disagree on predictions. Their concerns though were based on solid intelligence data on Iraq. First, stung by the destruction of its nuclear plant by Israeli aircraft in 1981, Iraq had gone on a multibillion-dollar spending binge and had built an extensive air defense force. It consisted of an integrated series of modern radars, communications centers, airfields, and fighter-interceptors. The American planners had a healthy respect for the system and called it "KARI," backwards for the French spelling of Iraq.

Second, historical data showed that 85 percent of all aircraft shot down were downed by AAA and shoulder-fired SAMs.[78] Intelligence data showed that KARI also included over 8,000 AAA weapons in both fixed and mobile sites. Additionally, the Iraqis had almost that many heat-seeking (infrared) SAMs. These two weapons combined were especially dangerous to helicopters.[79]

Before the coalition aircraft could have much of an impact against Iraq, they had to deal with KARI. Phase I of the air campaign was intended, in part, to eliminate Iraq's ability to shoot down our aircraft. Traditionally, this was done through a "rollback" effort designed to systematically destroy enemy air defenses. But the Black Hole desired to try a new approach with some tactical innovations. Using the latest in stealth and cruise missile technologies, they proposed attacking the heart of KARI at the beginning. A fleet of F-117s and cruise missiles would strike into the heart of Iraq and destroy their command and control centers. Then, when the various sectors had been broken down into independent operations, they would destroy selected radar sites, which would create a corridor for waves of conventional strike aircraft to enter Iraq and begin hitting the designated strategic targets. As they did, they would be supported with electronic jamming aircraft, accompanied by F-4G "Wild Weasels" and F-18s, both firing HARMs that would then find and destroy the individual enemy radars.[80]

To do the sophisticated and highly technical planning for the "electronic war," Horner brought over an expert in that business, Brig Gen Larry Henry. Henry was a career electronic warfare specialist and had acquired the nom de guerre of "Poobah." He set out to determine the best way to strip Iraq of its ability to defend itself from the onslaught of the coalition air effort. He and his small but highly specialized planning team began to look for ways to blow holes through the enemy radar picket lines along the Saudi-Iraqi border.

Task Force Normandy

Thinking along the same lines, Capt Randy O'Boyle made a novel suggestion. Intelligence showed that the Iraqis had placed early warning radars as close as one mile to the Saudi border. He suggested putting in Army SOF teams to attack and destroy the sites. SOCCENT worked up a plan to accomplish that. But to be successful, the teams needed precision navigation gear. None was available, so the mission was assigned instead to the MH-53s of the 20th SOS.

Meanwhile the Iraqis moved the radars 20 miles back into Iraq and hardened the sites. Noting this change, Colonel Comer did not feel that his MH-53s would be able to destroy the sites without being detected and reported to Baghdad. He discussed it with Col Ben Orrell. Orrell suggested that Comer bring in some Army Apache helicopters to attack the sites with Hellfire missiles and rockets. The MH-53s could use their GPS to get them to the correct firing positions because they were much more accurate and reliable than the Doppler systems on the Army helicopters.

They took the suggestion to Colonel Johnson at SOCCENT. He backed it and took the plan to General Schwarzkopf, who approved the use of Apaches from the 101st Air Assault Division and cleared them to begin training. Comer then met with LTC Dick Cody, commander of the 1st Battalion of the 101st Aviation Regiment, and they began planning the mission. Meanwhile, the Iraqis shut down one of the sites. Comer and Cody then developed a simple plan where two MH-53s would lead a small team of Apaches to each remaining site and provide combat recovery support. The two units trained for the mission through the fall. They also received permission to live-fire six Hellfire missiles in the Saudi desert.[81]

In late October Colonel Gray personally briefed General Schwarzkopf that the joint team was ready to execute its mission. When Gray assured him that the mission would be 100 percent successful, Schwarzkopf replied, "Okay, Colonel, then you get to start the war."[82] The joint team held a final rehearsal in January, and it went perfectly.

Poobah's Party

Another approach to the problem has come to be known as "Poobah's Party," so named for the man who planned it. General Henry, using detailed intelligence of the Iraqi air defenses and an intense study of Israeli tactics against the Syrians, developed a plan to bloody the noses of the Iraqi gunners and missile crews and dampen their enthusiasm for shooting down coalition aircraft.

His plan was simple in concept but challenging in execution, for it depended on a closely orchestrated series of events. First, stealth F-117s and BGM-109 Tomahawk cruise missiles launched from US Navy ships in the Gulf would saturate targets around Baghdad and thoroughly intimidate the Iraqi defenders. Some of the cruise missiles would carry special carbon filaments that would lie down over electrical lines and short them out causing all manner of unpredictable results.

With the enemy confused but fully alert and cocked to fight, he would then flood the skies over Baghdad and western Iraq with BQM-74 unmanned drone aircraft. Seeing what appeared to be such easy targets, the Iraqi radars would go to full power and radiate their signals. These signals would be easy to identify and locate. Then Henry would send in a wall of F-4G Wild Weasels and F-18s loaded with specially designed antiradiation missiles tuned to home in on those active radars and destroy them.

The plan would, to the limits of available resources, provide every package of aircraft going in to attack any heavily defended site with jamming support from EF-111s or the Navy's EA-6Bs, F-4Gs, or F-18s. They would attack SAM sites or radar-controlled guns that might threaten any coalition aircraft.

Additionally, Henry was betting that when the AAA gunners saw what happened to the SAM sites during Poobah's Party they would not turn on their tracking radars. This would mean that they would be much less effective.

To take advantage of this turn of events, General Horner directed his aircraft to stay above 10,000 feet. Since traditionally, most aircraft shot down are lost to guns, this meant that fewer aircraft would be lost.[83]

This was a radical departure for the Air Force. For the last 15 years, it had based its tactics upon low-level operations. Those tactics were based on the assumption that in facing the Soviet Union, the predominant threat would be the preponderance of SAMs and radar-controlled guns. History suggested that the only way to maintain any effectiveness in such an environment was to operate at low level and use terrain masking.

Years of training, though, made some resistant to change. Several of the fighter wing commanders requested a waiver to this policy. They argued that they needed to operate at low altitude because their tactics, specialized navigation/targeting equipment, and weaponry were optimized for low-altitude operations. Based on these arguments General Horner allowed exceptions to the F-15Es, the F-111s, and British, Italian, and Saudi Tornadoes. Most pilots though could clearly understand the need for the change and trained for the medium-altitude tactics.[84]

Of course fewer aircraft shot down would mean fewer rescues. Additionally, Army and Marine ground units stood prepared to support the air operations. Some artillery units did fire on enemy AAA and radar sites during the air campaign. At one point the US Army VII Corps had its artillery fire an ATACMS surface-to-surface missile to completely destroy an SA-2 missile site.[85]

At the 20th SOS, Colonel Comer and his men attacked the rescue problem from a more practical aspect. He said, "When we looked at the air war . . . we came to the realization that there were going to be more shoot downs than we could shake a stick at. And we would really have liked for none of those shoot downs to be us, so how do we do that, and how do we execute the mission successfully and get out of there? We had all kinds of think tank kind of sessions where we said, 'What do we tell the fighter guys to do if they are out there on the ground? How do they signal us?'"[86]

At the squadron level, their concerns were more practical. Grandiose plans were wonderful, but they had to actually fly the missions. And missions could succeed or fail on the smallest of details.

Proven Force

Within days of the initial deployment of US forces to the Gulf regions, senior US commanders in Europe began conceptualizing a force that could open a second front against Iraq. As the war plan against Iraq continued to mature, planning was going on in Europe to design an aerial force that would strike Iraq from the north. Based in Turkey, it would be multifunctional. The concept for the plan actually originated with tactical planners in the 52d Tactical Fighter Wing (TFW) at Spangdahlem Air Base, Germany.

Throughout the fall of 1990, as the various elements of US and coalition power came together in the Gulf, a small group of aircrewmen designed a notional package of aircraft and support units that could quickly deploy to air bases in Turkey and conduct sustained operations against the northern portion of Iraq. This force would be designed to attack the same array of targets as the forces in the south. It would also have a robust electronic warfare capability and a special operations task force included for combat recovery. These special operations forces would be able to extend combat recovery coverage across the northern half of the country.

This plan was refined and passed up the chain of command to General Powell. He was generally supportive and had the plan briefed to the Turkish General Staff. But the plan needed the approval of the Turkish government, and that was problematic because of the political realities in the region. The Turkish government and its leader, Pres. Turgut Özal, supported the United Nations and US actions and had taken several political and economic steps in parallel. But allowing combat sorties to be launched against a neighbor was another matter.[87]

Regardless the European Command (EUCOM) progressed to detailed planning. Since this was in support of CENTCOM, that command was designated as the "supported" force and EUCOM was the "supporting" force. Planners at EUCOM designed a joint task force named Proven Force. It would be commanded by USAF Maj Gen James Jamerson. From a CSAR perspective, this was fortuitous because Jamerson was a career fighter pilot and had flown a tour in Southeast Asia in A-1s. He had participated in or led many rescue missions and understood the mission intuitively.

By late December a formidable array of fighter, attack, and support aircraft, equipment, and personnel had been assembled at Incirlik Air Base in eastern Turkey. All aircraft remained under the control of their parent units, but plans were published to quickly establish a provisional wing and initiate combat operations as soon as political permission was granted. In support, a special operations sub-element made up of troops from the 39th SOW and the 10th Special Forces Group would assume combat recovery alert with MH-53s and support MC-130s at Batman Air Base.[88]

But the implementation of the plan still needed Turkish approval. And as combat approached, it did not seem to be forthcoming.[89]

Eastern Exit

Ready to perform the combat recovery mission and the opening night raid, the 20th SOS was now alerted for a mission in a totally different area. Serious internal unrest was sweeping the nation of Somalia on the eastern coast of Africa. CENTCOM alerted the unit to be ready to fly there on 10 January to evacuate US and coalition personnel. This threw the unit into chaos since it was now faced with the possibility of having three simultaneous missions in, literally, different parts of the world.

Fortunately, the orders were changed. CENTCOM decided to send a Marine expeditionary unit instead. Launching a rescue force of 55 marines and Navy SEALs off of the USS *Trenton*, they recovered 281 Americans and other nationals from the US Embassy. Released from the Somalia tasking, the 20th SOS was able to refocus on the Task Force Normandy mission and its combat recovery alert.[90]

Eve of Battle

On the eve of battle, the CSAR capability of CENTCOM was extensive. General Horner was the overall SC. Through his JRCC he could coordinate with other components for CSAR. The JRCC, working through the TACC, had the authority to task any asset under the general's control. This could be anything from fighters to communications to reconnaissance, but he had no dedicated combat recovery forces assigned, and he could not directly task other components.

Each service or ally had unique forces or capabilities for CSAR. Each had its own RCC and was expected to prosecute its own recovery missions to the maximum extent possible. Situations beyond its abilities or control had to be passed to the JRCC.

The US Army and Marines each had hundreds of helicopters that could quickly respond. The British, French, and Saudi forces had helicopters and other support aircraft available. The US Air Force had none.

SOCCENT had the most capable recovery forces in the theater, but it retained OPCON of its forces even though General Horner was the SC. It had 37 Air Force, Navy, and Army helicopters and eight Air Force tankers plus US Army Special Forces teams and US Navy SEAL teams. Those assets were arrayed over a number of airfields from ArAr to Ras Al Mishab and on ships in the Gulf ready to answer the call.[91]

Additionally, SOCCENT anticipated that if Turkey approved the deployment and activation of the Proven Force task force, they would have recovery assets available to support them from the north. Surveying the overall operation from his perspective, Col Ben Orrell at SOCCENT said, "We were pretty well prepared."[92]

Notes

1. Andrew Leyden, *Gulf War Debriefing Book* [hereafter *GWDB*] (Grants Pass, OR: Hellgate Press, 1997), 104.
2. Williamson Murray with Wayne Thompson, *Air War in the Persian Gulf* (Baltimore, MD: Nautical and Aviation Publishing Company of America, 1996), 32.
3. Michael R. Rip and James M. Hasik, *The Precision Revolution: GPS and the Future of Aerial Warfare* (Annapolis, MD: Naval Institute Press, 2002), 132.
4. GEN H. Norman Schwarzkopf with Peter Petre, *It Doesn't Take a Hero* (New York: Bantam Books, 1992), 310.
5. LTC Pete Harvell, interview by the author, 29 January 2002.
6. Leyden, *GWDB*, 106.
7. Murray with Thompson, *Air War in the Persian Gulf*, 10.
8. *Desert Shield/Desert Storm: Air Force Special Operations Command (AFSOC) in the Gulf War* [hereafter *AFSOC in the GW*] (Hurlburt Field, FL: History Office, Air Force Special Operations Command, 1992), 13.
9. Col Ben Orrell, 1st Special Operations Wing (SOW), 16th SOW History Office, Hurlburt Field, FL, interview by the author, undated, [hereafter Orrell, 1st SOW interview].
10. AFSOC History Office, *AFSOC in the GW*, 14.
11. Maj Joseph J. Falzone, *Combat Search and Rescue: CSEL Enhancements for Winning Air Campaigns* (Maxwell AFB, AL: Air University Press, December 1994), 55.
12. Lt Col Larry Helgeson, interview by the author, 2 September 2002.
13. Brig Gen Dale Stovall, interview by the author, 3 September 2001.
14. Michael R. Gordon and Gen Bernard E. Trainor, *The Generals' War: The Inside Story of the Conflict in the Gulf* (New York: Little, Brown and Co., 1994), 250.
15. Stovall, interview.
16. Harvell, interview.
17. Stovall, interview.
18. Col George Gray, interview by the author, 3 May 2001.
19. Lt Col Richard Comer, interview by the author, 19 July 2000.
20. Tom Clancy with Gen Carl Stiner, *Shadow Warriors* (New York: G. P. Putnam's Sons, 2002), 449; and Susan L. Marquis, *Unconventional Warfare* (Washington, DC: Brookings Institution Press, 1997), 238.
21. Lt Col Joe E. Tyner, *AF Rescue and AFSOF: Overcoming Past Rivalries for Combat Rescue Partnership Tomorrow*, National Defense Fellows Program (Maxwell AFB, AL: Headquarters US Air Force and Air University, undated), 37.
22. Maj Mark Tucker, interview by the author, 6 June 2002. Many rescue troops who were interviewed for this book expressed the same sense of betrayal. They wanted to go. The bitterness was intensified by subsequent events in Bosnia and Serbia in the late 1990s, and these feelings still exist. One career rescueman, Col Ken Pribyla, served as the director of operations at Air Rescue Service after Desert Storm. He remembered that, "The frustration was great throughout Rescue as I was told numerous times that there

were a number of options available from single units to a merging of units, both active and reserve component, that could have provided the needed Combat Search and Rescue (CSAR) units. Conventional wisdom said that since Rescue did not make the war, its requirements remained mostly unknown. Additionally, special operations forces became the visible force of choice when rescue was needed in any theater. This paradigm became the ultimate catch-22 for ARS forces. You couldn't get to the war/fight if you hadn't been to the war/fight. The pattern continued well into the decade and created a considerable amount of confusion as to which organization was truly the Air Force rescue force."
E-mail from Col Ken Pribyla, USAF retired, to author, 8 September 2002.

23. Gordon and Trainor, *Generals' War*, 243.
24. Gray, interview.
25. Al Santoli, *Leading the Way, How Vietnam Veterans Rebuilt the U.S. Military: An Oral History* (New York: Ballantine Books, 1993), 206.
26. Harvell, interview.
27. Comer, interview.
28. Capt Tom Trask, interview by the author, 16 February 2000.
29. Perry D. Jamieson, *Lucrative Targets: The U.S. Air Force in the Kuwaiti Theater of Operations* (Washington, DC: Air Force History and Museums Program, 2001), 20.
30. Murray with Thompson, *Air War in the Persian Gulf*, 22.
31. *History of the Air Force Special Operations Command*, Hurlburt Field, FL: [hereafter *History of AFSOC*], 1 January 1990–31 December 1991, 37.
32. Capt Randy O'Boyle, interview by author, 20 March 2000; and History of AFSOC, 26.
33. Brig Gen Buster C. Glosson, interview by the author, 25 September 2002.
34. Appendix 6 to Annex C to USCINCCENT OPLAN 1002-90, 18 July 1990.
35. Ibid.
36. Appendix 6 to Annex C to COMUSCENTAF OPLAN 1021-90, 30 March 1990.
37. *History of AFSOC*, 27.
38. Lt Col Joe Hampton, interview by the author, 12 March 2000; and Joint Review, JCS.
39. Hampton, interview.
40. *History of AFSOC*, 37.
41. Comer, interview.
42. *Operation Desert Shield Combat Search and Rescue Plan*, USCENTCOM/JRCC, 1 November 1990.
43. Hampton, interview.
44. *Operation Desert Shield Combat Search and Rescue Plan*, USCENTCOM/JRCC, 1 November 1990; and Gray, interview.
45. Harvell, interview.
46. Capt John Steube, interview by the author, 9 January 2002.
47. *CNA Desert Storm Reconstruction Report, Volume II: Strike Warfare*, October 1991, 5–52. [hereafter *CNA*]
48. *History of AFSOC*, 374.

49. Tyner, *AF Rescue and AFSOF*, 40; and Gray, interview.
50. Gray, interview.
51. Gordon and Trainor, *Generals' War*, 250.
52. Gray, interview; and *US Special Operations Command History*, MacDill AFB, FL, HQ USSOCOM/SOCS-HO, November 1999, 36. [hereafter *SOCCOM History*]
53. *History of AFSOC*, 28.
54. Ibid.
55. Gordon and Trainor, *Generals' War*, 251.
56. O'Boyle, interview.
57. Comer, interview.
58. Hampton, interview.
59. Harvell, interview.
60. Detachment Summary Report, *Helicopter Combat Support Special Squadron 4, (HCS-4), Desert Shield/Desert Storm*, 9 December 1990–20 March 1991. Includes attached aircraft and personnel from HCS-5.
61. History of the 919th Special Operations Group, 1 January 1990–31 December 1991.
62. 3d/160th SOAR Desert Storm AAR, file 525.
63. CDR Patrick Sharrett, USN; Lt Col Joseph Wysocki and Capt Gary Freeland, USAF; CPT Scott Netherland, US Army; and Donald Brown, "GPS Performance: An Initial Assessment," Fairfax, VA, September 1991. Paper procured from the Institute of Navigation, 395.
64. Ibid., 396.
65. For a full explanation of GPS in Desert Storm, see chap. 5 of Rip and Hasik, *Precision Revolution*.
66. Tom Clancy with Gen Chuck Horner, *Every Man a Tiger* (New York: G. P. Putnam's Sons, 1999), 274.
67. Ibid., 271.
68. *Gulf War Air Power Survey (GWAPS) Summary Report*, vol. 5 (Washington, DC: Government Printing Office, 1993), 180.
69. Ibid.
70. Trask, interview.
71. Comer, interview.
72. Murray with Thompson, *Air War in the Persian Gulf*, 58.
73. *Final Report to Congress: Conduct of the Persian Gulf War*, Department of Defense, April 1992, 120.
74. Glosson, interview.
75. Clancy with Horner, *Every Man a Tiger*, 559.
76. Col Ali Abdul-Lateef Khalifouh, Kuwait: Air Force, retired, *Kuwaiti Resistance as Revealed by Iraqi Documents*, Center for Research and Studies on Kuwait, 1994, 36.
77. Gordon and Trainor, *Generals' War*, 90.
78. Ibid., 111.
79. Murray with Thompson, *Air War in the Persian Gulf*, 70.
80. Ibid., 43.

81. First Special Operations Wing (SOW) History, 1 January–31 December 1991, and support documents, 121.
82. Gray, interview.
83. Clancy with Horner, *Every Man a Tiger*, 351.
84. Murray with Thompson, *Air War in the Persian Gulf*, 137.
85. Maj Mason Carpenter, *Joint Operations in the Gulf War* (Maxwell AFB, AL: School of Advanced Airpower Studies, Air University, February 1995), 57.
86. Comer, interview.
87. Joint Task Force Proven Force, http://www.fas.org/man/dod-101/ops/proven_force.htm.
88. Clancy with Stiner, *Shadow Warriors*, 411.
89. Joint Task Force Proven Force.
90. First SOW History, 122.
91. General briefing, CSAR file, SOCCENT, undated.
92. Orrell, interview; and 1st SOW, interview.

Chapter 4

Desert Storm Week One: 17–23 January 1991

We were honor bound to do our best to rescue them if they were shot down.

—Lt Gen Chuck Horner

You just can't go trundling into some place in a high-threat environment without knowing exactly where the guy is.

—Capt Randy O'Boyle

As preparations for war were proceeding, national leaders and diplomats made last-minute efforts to forestall what appeared to be inevitable. In Saudi Arabia events were taking place at breakneck speed as final changes were made to the overall war plan, and units redeployed to the best tactical locations as a prelude to combat.

Were the aircrews ready? Many of the older pilots had flown combat and generally knew what to expect. Among the younger guys, though, there were the inevitable doubts.

Said young A-10 pilot Capt Lee Wyatt, "You always wonder if you're going to be ready for the real test. I had just reread *Thud Ridge*, and I thought about those guys that flew through the hell in North Vietnam and I wondered if I was ready to be shot at."[1]

His unit was visited by Lt Col Greg Wilson, a member of the 706th Tactical Fighter Squadron (TFS), an Air Force Reserve unit activated for the war. Wilson had flown as a forward air controller in Southeast Asia and had extensive combat experience. He laid it out for the young guys. "You will have doubts, but you are ready for it. You're prepared, you're trained for it, and you'll do what you are trained to do."[2] The pilots were strengthened by his words.

Brig Gen Buster Glosson visited the units to fire up the aircrews. Based on his experiences in Southeast Asia, he felt an obligation to look each of them in the eye and tell them why they were fighting. He told them that the nation was behind them. He explained to them that what they were doing was important and talked them through the initial air campaign plan. He cautioned them to be careful and pointed out that the fighting could last awhile and that resources needed to be husbanded. He said, "There is not a . . . thing in Iraq worth you dying for."[3]

It made "the hair stand up on the back of your neck," according to Capt Bill Andrews of the 10th TFS at Abu Dhabi.[4] Several of the pilots recalled that General Glosson also made an appealing and specific promise. As per Saudi law, the troops had not been allowed to have alcoholic beverages in Saudi Arabia. Glosson promised them all that after the shooting was over, he would revisit all the bases with a planeload of cold beer. That cheered them up.[5]

The British pilots' approach was a bit more pragmatic. One remembered, "The brief that I'd been given was that we were there to use our assets as efficiently as possible, without losing aircraft or lives. I think our outlook was slightly different to how it would have been had we been defending British territory. We were there to do a job and we were quite happy to do it, but none of us intended to die in the process."[6]

As combat approached, some of the wing commanders wanted to know what procedures the flyers should follow for CSAR. Glosson had Randy O'Boyle prepare a briefing for the units on what to expect. O'Boyle conferred with Colonel Comer on what to say and then split the duty with Capt Corby Martin and Capt Tim Minish. O'Boyle described what he told the flyers: "I said, 'This is how we are going to try to come and get you.' And I talked all about the SEALs, their desert vehicles, and . . . some of the things that we were going to use . . . and basically gave them the plan which was a real rah-rah speech but, I thought, realistic. I don't think that I overstated it. Before I got back, one wing commander called General Glosson and said, 'That was . . . hot. That was what the guys needed to hear.'"[7]

One pilot, a Marine F-18 jock, recalled, "Representatives from the Air Force MH-53 helicopter units who would do the actual rescue work came to brief us. These folks and their equipment were impressive. Much of the briefing was classified, but we were comforted somewhat by the material they were putting out. One interesting point was made: there would be no repeats of the rescues from the Vietnam War, where all stops were pulled. There, all-out efforts were made without much regard to the cost in lives or machinery, just to rescue a single pilot. Neither the people nor the equipment could be afforded now."[8]

Another Marine pilot remembered a more severe message. His commander told their unit, "Let me tell you something. When you get shot down, you are on your own. If you go out with that attitude, you're going to be okay. But when you get out of your aircraft, it is one million BC. Expect that no one is going to help you . . . and you're going to be okay."[9]

One unit seemed to have heard a different message. When Glosson visited the 4th Wing at Al Kharj, one young pilot remembered him saying that, "If you get shot down, you [won't] have to spend the night. If it's a night sortie, we'll be in the next day to get you, but you are not going to spend 24 hours on the ground."[10]

Glosson himself remembered saying more than once, "Don't worry about it. We'll pick you up if we have to stack helicopters on top of each other and get them all shot down. But we are not going to leave you out there."[11]

The crews were also given escape-and-evasion briefs. They were checked out in the use of all of their survival equipment and brought up-to-date on the theater CSAR procedures. There were some items of concern. The terrain over which they would be flying was, for the most part, dry and flat. There were few places to really hide. The area was teeming with snakes, spiders, and all forms of biting creatures, and the Bedouin tribes roamed freely and randomly. They had been promised big rewards by Saddam if they captured any coalition personnel.

Aboard the USS *Saratoga* in the Red Sea Battle Group, LT Rick Scudder was a helicopter pilot assigned to the staff of the air wing commander. One of his many duties was CSAR coordination. He remembered the visit by Captain O'Boyle.

Randy came out to explain two new features of the air tasking order:

1. where to go each day to find information germane to CSAR—the safe areas, other time-critical and perishable information, information that was subject to change and

what the procedures would be should there be suspected compromise, etc.; All the "what-if" stuff.
2. the importance of SPINS [special instructions for CSAR]. O'Boyle reviewed them in detail and said, [paraphrasing] "Every pilot on the ship has to commit this absolutely, positively to memory because obviously you can't take it with you. We really don't want you condensing it into kneeboard cards. You just have to know this stuff. You have to know before you go."[12]

According to Scudder, O'Boyle laid it out for the flyers and was well received. He spoke to nearly every crewmember in the wing. He was also realistic about the chances of rescue. O'Boyle pointed out that opportunistic missions in a medium- or high-threat environment were very problematic and that helicopters were very vulnerable in daylight, especially in a wide-open area like Iraq. Scudder said of O'Boyle, "There wasn't any blanket promise: 'You go down, we're inbound right away' kind of thing."[13]

One young *Saratoga* flyer was not impressed with the briefings. LT Jeff Zaun, a bombardier/navigator in A-6s with VA-35 said, "We had plenty of CSAR briefings. The difficulty was that there was a lot of new information. We were focused on flying our mission. You want to be professional and make sure you have all the CSAR stuff you need. But it's like, somebody is coming to town with a new project and you think—look, I have a place to bomb and you are telling me all this stuff about CSAR?"[14] Simultaneously, a team of personnel from life support, rescue units, and the JRCC traveled to all the units in Turkey and briefed them on CSAR procedures and SPINS preparatory to their being authorized to fly against Iraq.

One pilot from the 23d Tactical Air Support Squadron, flying the OA-10 as a forward air controller, had low expectations of rescue. He and his cohorts would be flying over Kuwait. He said in looking at Kuwait that, "We were on a pool table out there. There were a couple of hundred thousand Iraqi troops. My impression was that if I was ever shot down, I was going to be a POW because there really wasn't any way to get me out. . . . Most guys were convinced that we were going to be grabbed immediately because in A-10s we were working right there on the border or 40–50 miles across. It was fairly densely populated with bad guys wherever we were working. I didn't think that they would have cleared a helicopter across the border just to get a guy down there. You would just lose the helicopter too."[15]

In Washington President Bush gave General Schwarzkopf the authority to set the actual date and time for initiating the attack, as long as it was as soon after 15 January as possible. Since the attack would initially be by air, Schwarzkopf deferred that decision to General Horner. Horner's planners recommended the early morning hours of 17 January because there would be no moonlight to illuminate the F-117s, which would be the first coalition aircraft over Baghdad.[16]

Finally, the TACC published the air tasking order or ATO. This document, produced daily, provided the detailed mission assignments for each of the flying units in each 24-hour period. It included the SPINS for CSAR. It was transmitted to all units and flown out to the aircraft carriers.

On 13 January US government representatives formally briefed Turkish President Özal on Proven Force. Two days later, after meetings with his government, he approved the deployment of the joint task force but hesitated to allow it to begin operations. The next day US Air Forces in Europe (USAFE) published the operational order and activated the 7440th Composite Wing (Provisional) as the overarching headquarters for the deployment. Most required aircraft were in place at Incirlik Air Base. They were quickly joined by many more from throughout Europe. Eventually, 24 F-16s, 18 F-111Es, 22 F-15Cs, 13 F-4G Wild Weasels, six RF-4s, six EF-111s, three AWACs, and 14 KC-135 tankers parked on the ramp at Incirlik. They were reinforced by a special operations task force which, when ordered, would deploy forward to the small airfield at Batman, Turkey, with seven MC-130s and seven MH-53J Pave Low IIIs from the 21st SOS and assume combat recovery alert.[17]

The task force was also supported with batteries of Patriot antiaircraft and antimissile missiles that were brought in to defend both Incirlik and Batman air bases. A large contingent of air defense aircraft from several NATO countries was deployed in eastern Turkey to counter any possible attack by the Iraqi Air Force, but they were not part of Proven Force.[18]

In the JRCC, Colonel Hampton was notified that Proven Force would have its own JRCC. He was confused. He had sat in on a meeting between SOCCENT and Proven Force representatives and knew that any special operations actions out of Turkey would be in support of SOCCENT. This suggested to him that for rescue, they would set up an RCC that would then respond to the CENTAF JRCC. Then he started receiving calls on his secure phone requesting CSAR data, but he did not have a good secure data communications link with them and could not pass classified information in a timely manner. Subsequently, the Proven Force RCC began to work more and more autonomously, like a JRCC. Joe Hampton and his crew scrambled to improve the communication links with the rescue forces in Turkey. Given their remoteness, it was obviously going to be a challenge to work with them if and when they got all the political approvals necessary to begin operations.[19]

On 16 January all SOCCENT units were notified that combat operations against the Iraqis would commence at 0300L (local time) on 17 January. With that, they moved several aircraft to Al Jouf and ArAr as per the Task Force Normandy plan and the CENTCOM CSAR plan.

On the 16th General Schwarzkopf visited the air planners in the Black Hole to get a final update on the air plan. Horner had Glosson and Deptula give the briefing. Methodically, they took Schwarzkopf through the intricate and complicated flow of aircraft. When Schwarzkopf heard that no B-52s were scheduled to hit the Republican Guard units in Kuwait the first day, he erupted. The general had a healthy respect for the B-52s because he had seen their devastating power in Vietnam. He demanded that they be refragged (rescheduled) to hit the Iraqi units from the beginning.

Glosson was taken aback. He knew that putting the bombers into such high-threat areas before the air defenses had been attacked put his aircrews at great risk. He envisioned a repeat of the B-52s shot down over Hanoi during the Christmas raids of 1972. Perhaps uncivilly, Glosson asked Schwarzkopf, "Tell me how many airmen you are willing to lose, and the air war commanders could redraft [their] plan to attack the Republican Guard in the first hours of the war."[20] Horner interceded at that point and led the commander into his office where they discussed the issue privately. After Schwarzkopf left, Glosson assigned reconnaissance units to closely watch the Republican

Guard for any unpredicted movement and left the air plan unchanged.[21]

As the air war began, morale among the aircrews was high. They were well trained and equipped and believed strongly in the mission—a belief that extended to the support troops. One B-52 pilot noted, "We were all pumped. We were even more pumped as we took off and saw all our maintenance troops lining the runways cheering and waving flags."[22]

The "execute" order made its way down through the units to the men who would prosecute the war. In the Red Sea, the task force commander, RADM George Gee, spoke to his sailors: "Gentlemen, [President] Bush has called upon us to do our duty, to liberate Kuwait, and that liberation is going to start tonight. [The American people] are going to be watching you. You need to do this well. This will be with you for the rest of your lives. You will remember this night forever, so you want to do the best job you possibly can because if you don't, you will regret it until you die."[23]

Listening to the admiral, many of the pilots could feel their throats tighten. The pilots aboard the *Saratoga* remembered getting ready for their first mission. "We sanitized our flight suits. That means that all nametags and patches were removed. We wear our dog tags. Personal weapons are carried."[24]

One day prior to combat, the *Saratoga* received a shipment of the new PRC-112 survival radios. Lieutenant Scudder received the gear for the wing. His units had deployed with PRC-90s, and he was happy to receive the more capable radios. Quickly he and his life support specialists read the instructions and learned how to properly key them for the crewmembers. But there were not enough for all. They would have to be rotated among the men as they flew. Additionally, Scudder decided to parcel one to each F-18 pilot and one to each crew of an A-6 or F-14.[25]

Briefings were hastily arranged to teach the crews how to use the radios. Preoccupied with last-minute details of the first strikes, several aircrewmen found it difficult to concentrate on the new radio. Several were also concerned that the radios did not fit well into the survival vests. One survival specialist noted, "There's a good chance if you eject, this thing's not staying in your pocket."[26]

Lieutenant Zaun of VA-35 recalled that, "The day before [the beginning of combat operations] they came in with the new radios. I basically said, 'No, I am not carrying a new radio. I never used it and don't know how it works. I am sure I could figure it out but I've got other things to do right now. So I am taking the PRC-90.'"[27]

The commander of the air wing on the *Saratoga*, CAPT Dean Hendrickson, also talked to his flyers. The veteran of aerial combat in Southeast Asia said, "You are going to come back. Then you're going to look around . . . and one of you won't be here."[28] That sent a shiver through the flyers and focused them on the stark reality of what they were about to do.

Over in the Gulf, the admiral aboard USS *Midway* broadcast a message to all hands, "We have received the execute order to strike targets in southeastern Iraq as part of what will be called Operation Desert Storm. At this point, I do not know how long this conflict will last. I do know that the men of the *Midway* will do their utmost to defeat this enemy with dispatch. In a few short hours, we will launch our first strike at the enemy. To the sailors and crew, keep alert and let's run the tightest of ships; and to the aviators, let's make each bomb count."[29]

One young flyer who heard the announcement recalled his feelings. "The adrenaline was flowing through our veins and we were ready to do what we had trained so hard to do; strike and win."[30]

The first waves of the attacks would be directed against the Iraqi air defenses. This was critical to the entire air campaign. "The systematic destruction of Saddam's eyes and ears was the one precondition necessary to implement the remaining phases of the air campaign," noted the executive officer of another of the Navy electronic warfare squadrons.[31] He and his squadron mates of VAQ-136 flew EA-6Bs and had the dangerous task of suppressing the Iraqi SAMs. They understood their mission perfectly. One young pilot, LT Sherman Baldwin, noted, "Our squadron's job was to ensure that every *Midway* aircraft that launched came back safely. If we did our job of suppressing Iraqi air defenses, their surface-to-air missiles would be ineffective, their fighters would not have the support of ground-based radars, and our guys would get in and out unharmed."[32]

As the aircrews were preparing to fly the first wave of strikes, the cruiser USS *San Jacinto* in the Red Sea and the destroyer USS *Paul F. Foster* in the Persian Gulf began to salvo 122 Tomahawk cruise missiles. Targeted against radar and SAM sites, they would lend support to the efforts being put forth to decrease the effectiveness of the Iraqi air defenders. Those stubby-winged, pilotless aircraft would be the first coalition aircraft to enter Iraqi airspace.[33]

For some aircrews, the war actually started early. CENTAF decided to run a "con"—a tactical deception mission. Its purpose was simple: to provide an excuse for closing the airspace along the Saudi-Iraqi border to civilian aircraft without tipping off the Iraqis. This was no small concern because several international jet routes between Europe and Asia ran through that area. The coalition planners needed a way to close the airspace for civil use so that the hundreds of fighter and attack aircraft slated to fly through there would not have to worry about possibly colliding with civilian airliners.[34]

Planners scripted a radio dialogue to be played by AWACS and several other aircraft to replicate a rescue mission for an F-117. Obviously, the Iraqis would monitor the operation. The planners hoped that the Iraqis would feel that the loss of such a "special" aircraft would explain all the unusual activity. It was to take place along the Saudi-Iraqi border northeast of ArAr. Two helicopters from HCS-4/5 were also participating. Unfortunately, the AWACS aircraft that was key to the operation—and whose crew knew the overall script—experienced a mechanical problem and had to return to base. No other crew had been briefed on the ploy, so the mission was cancelled. The HCS-4/5 birds were reassigned to backup combat rescue duty for the Task Force Normandy operation.[35]

The JRCC Log

As the overall CSAR coordinator for CENTCOM throughout its area of operations, the JRCC kept a running log of all incidents and missions. Whenever a suspect incident occurred, the controllers would assign it an incident number (I###) so that it could be tracked. An incident would be declared anytime there was any indication that an aircraft was in distress anywhere in the designated area of responsibility. This could be anything—an emergency beeper activation (called an emergency location transmitter/ELT), a loss of radio communications or flight following with an aircraft, an aircraft disappearing from radar contact, an aircraft squawking the emergency code, or an actual report of a downed aircraft from any source. All of the incident declarations were based on one or more of these factors. Mission numbers (M###) were applied once an effort was begun to dispatch a rescue mission.

17 January

Combat operations in Operation Desert Storm commenced on 17 January (H hour was 0000Z) and lasted until the ceasefire at 0500Z on 28 February. The JRCC log that begins below with Task Force Normandy provides an excellent tool for reviewing the CSAR events of Desert Storm. Prior to the ceasefire, there would be a total of 89 logged incidents. Since many were false squawks or beacon activations, they are not mentioned.

(Note: At the time of Desert Shield/Desert Storm, Iraq/Kuwait local time (L) was ZULU + 3 [i.e., Greenwich Mean Time plus three hours]. Saudi Arabia local time was ZULU + 3:30. All times in the log extracts that follow are referenced to ZULU time.)

Task Force Normandy

The time over target (TOT) for the helicopters was 0238L or 2338Z. The time was carefully calculated so that the huge wave of aircraft set to fly through the hole could then take maximum advantage of the chaos being caused by the first F-117 strike, cruise missiles, and the drones being launched (Poobah's Party).

At 0100L, the group of helicopters now called Task Force Normandy lifted off, led by the 20th SOS commander, Lt Col Rich Comer, who flew as one of the copilots. The aircraft commanders were Capt Corby Martin and Capt Ben Pulsifer. The second section was flown by Capt Michael Kingsley and Maj Robert Leonik. Each flight of two MH-53s led four AH-64s.[36] Trailing in combat rescue support were two HH-60s from HCS-4/5, led by

Photo courtesy of Bob Reed, HCS-4

The Apaches of Task Force Normandy were among the first manned aircraft to enter Iraq, led by MH-53s from the 20th SOS.

their skipper, CDR Neil Kinnear, from the cancelled deception operation, and two MH-60s from the 55th SOS.[37]

As the Task Force Normandy helicopters crossed into Iraq, they observed some small arms tracers, but they were inaccurate and of no consequence. Guided by their GPS units, the crews on the Pave Lows flew to the prebriefed drop-off points, where they threw out a bunch of green chemical sticks and then turned south. As they departed, the Apaches slowly passed over the chemical lights and, using prebriefed coordinates, updated their Doppler navigational systems for the final 10-mile run to their individual targets.

Slipping through the clear, dark night, they pulled into firing position exactly 90 seconds early. The gunners could see the sites. They matched the intelligence pictures that they had been shown. They could also see enemy troops around them. Suddenly the lights began to go off. One of the pilots mused, "I think they know we are here." Thirty seconds prior, the Apache crews turned on their ranging lasers. At exactly 0238L, they began firing their Hellfire missiles. Twenty seconds later, the deadly weapons began to detonate against the structures. The generators were first, then the command bunkers, and finally, the radar dishes themselves. The enemy soldiers died in the melee.

Once the Hellfires were totally expended, the helicopters flew toward the sites and salvoed their rockets. Two thousand meters from the sites, they opened up with their 30 mm chain guns and riddled what remained of the compounds with every bullet they had. Four minutes after it started, it was over. The Apaches turned south, rejoined with the Pave Lows, and headed home.[38]

Outbound, they sent a message to SOCCENT headquarters reporting their complete success. "SOF targets destroyed," Colonel Johnson personally reported to General Schwarzkopf's command center.[39]

The combination of the Pave Lows and Apaches had worked perfectly. One of the Pave Low pilots, Capt Mike Kingsley noted their contribution to the operation when he said, "We were the logical choice because we have an advanced navigational system. We are very accurate. The Apaches do not have a sophisticated navigational system. They had the confidence in us to lead them in so that when it was time to destroy these radar sites they were fresh and ready to go."[40]

Orbiting at the border were the two HH-60s of HCS-4/5. They had not been needed for rescue. Looking south the Navy crews saw above them a wave of aircraft heading north at high speed. Commander Kinnear was awestruck by the sight. There appeared to be an endless stream of flashing aircraft beacons. It was the first wave, at the right place at the right time. As they passed over the Navy helicopter crews, they switched off their lights and disappeared into the black night.[41]

The first aircraft through were not fighters or attack aircraft, but EF-111 jammers. They would overwhelm the Iraqi radars with electrons and blind them for the critical minutes that the attack aircraft needed to get in and hit their targets. Leading the first three-ship flight from the 390th Electronic Combat Squadron was the commander, Lt Col Dennis Hardziej. Passing the border, they dropped down to penetrate Iraq at low level.[42]

One of the attack pilots remembered crossing the line. In the midst of a swirling mass of aircraft he noted, "The number of targets on the [on board] radar increased dramatically . . . this was a mass raid from all sides, all levels, heights, everything, running up towards the border, aiming to cross together in a

DESERT STORM WEEK ONE

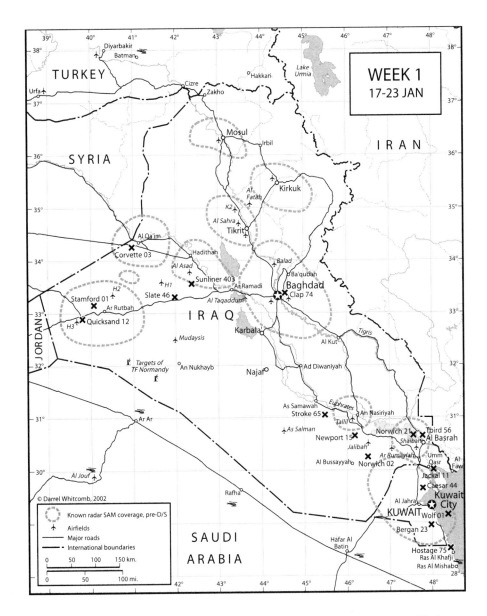

blanket push. They reached a certain point, switched off their IFFs [identification, friend or foe] and in they went."[43]

It was a perfectly clear night. The troops at ArAr watched the show. One helicopter pilot said, "I turned back to the north and was looking over where H2 and H3 [Iraqi airfields] were and off to the east toward Baghdad . . . when all hell broke loose. With the naked eye, you could see the tracers coming up. If you put on your [night-vision] goggles you could really see them. There were tracers and crap flying everywhere. We were sitting there watching this and you could see the flashes of light everywhere. Directly north about 60 miles, you could see a pattern of bombs going off. A B-52 strike."[44]

Poobah's Party

The Iraqis had quite a time as they downed most of the decoys launched by the Air Force and Navy. The drones appeared as manned aircraft on the enemy radars and soaked up an estimated nine missiles per drone or TALD (tactical air-launched decoy).[45] The F-4Gs were right behind them, and their antiradiation missiles destroyed dozens of radars and, undoubtedly, their crews as well. The desired effect was achieved. Within two days the enemy's use of its tactical radars had dropped by 80 percent.[46] The enemy developed a deep respect for the F-4Gs. They continued to fire SAMs, but their radars were not turned on. Without radar guidance, the missiles were just bigger bullets and much easier to evade.

Colonel Hardziej led his flight of three EF-111s to their orbit point west of Baghdad without F-15 escorts. They had a 10-minute window during which they had to radiate toward several sites in the Baghdad area as waves of F-117s attacked the most dangerous targets. Watching over them, the AWACS crews began to call warnings that "bandits" (Iraqi fighters) were airborne and hunting for vulnerable coalition aircraft.

Finishing his mission, Hardziej turned his flight south. He sensed a presence and saw an Iraqi fighter with a bright spotlight planted a few feet off his right wing. He tried to take evasive action. As he was approaching the Saudi border, the interceptor fired an air-to-air missile that missed and exploded in the desert. Hardziej and his other two aircraft safely returned to Taif. They told others about their run-in with the Iraqi bandit and the spotlight.[47]

As the air campaign began to sweep across Iraq and Kuwait, things were beginning to get busy in the JRCC. Some of the activity was just the chaos of war. Any indication of a downed aircraft though, had to be checked out. As operations began, they did not have long to wait.

I001

This was an emergency squawk by an airborne aircraft. The pilot was contacted by the radar controllers and confirmed that he, his crew, and aircraft were all right. It was a non-event, but had to go into the JRCC log.[48]

***

Some GPS-equipped MH-47s from the 160th Special Operations Aviation Regiment (SOAR) were launched with specialists from the 1720th Special Tactics Group (STG) to mark the actual location of the breech in the Iraqi radar line for the fighters. They were to lay down special reflectors that could be identified by the radars on the coalition aircraft. Moving along the border, they put down 11 devices to mark the passage. Unfortunately, several of the reflectors were dropped near Bedouin camps. After the helicopter left the area, the natives either stole or destroyed a number of them. One of the aircraft, Python 50, had to make a second sortie to replace the pilfered deflectors on alternate sites. Their mission was supposed to be completed by early afternoon, but with the second trip, the crew did not complete the mission until late evening. Then they had to quickly return to their base at KKMC to reconfigure the aircraft for combat rescue alert duties.[49]

I003 – Python 50/MH-47, 160th SOAR

Delayed by the problems with delivering the radar reflectors, this aircraft was not ready for launch until 0200L. Then the crew of CWO3 Russ Hunter, CWO2 Dan Folse, and CWO3 Don Harward departed for their alert site at Rafha.[50] They had their unit operations officer, Capt Eric Peterson, and several Army and Air Force enlisted troops on board. Flying northwest they leveled off at 100 feet above the ground and flew with their lights off. The darkness and uneven terrain made the flying challenging.

About 15 miles from Rafha one of the enlisted troops spotted an SA-7 missile being fired at the aircraft by what was later determined to be a small team of Iraqi special forces. When the soldier called a warning on the intercom, Hunter took immediate evasive action. Two more missiles were fired. Hunter's maneuvering and deployment of flares threw off the missiles. Disoriented by the light of the flares, exploding missiles, and abrupt maneuvers, Hunter allowed the aircraft to descend. It stuck the ground, and two of the aircraft's landing gear were severely damaged. Fortunately, none of the personnel on board was injured. Quickly recovering control, Hunter flew directly to Rafha and set the aircraft down on a bed of mattresses. Following emergency repairs, the crew flew back to KKMC after sunrise for complete repairs.[51]

***

Twenty minutes after Task Force Normandy destroyed the radars, a flight of F-18s from the USS *Saratoga* entered Iraq. As they crossed the border the pilots turned off their IFF and lights. Tasked to perform SAM suppression, the F-18s were carrying HARMs in support of a squadron of A-6s that was fragged to hit the airfield at Al Taqaddum, 25 miles west of Baghdad. They also had EA-6Bs and F-14s in support. All of this was coordinated with a large strike force coming in from the south. Below and ahead of them were the BQM-74 drones and TALDs launched as part of Poobah's Party. As they approached the target, the enemy reaction was intense. Several SAM sites were active. Iraqi MiG-25 interceptors launched, and the airfield was covered with AAA.[52]

CDR Bob Stumpf in the strike package remembered the scene. Seeing the glow stretching across the horizon and the "golden BBs" streaming up from the ground, he thought, "We didn't really expect this, but pretty much as soon as we entered Iraqi airspace, there were missiles and AAA." Stumpf had flown in the 1986 raid on Libya. He knew what ground fire looked like. But that did not compare in lethality to this. He said to himself, "I think I am going to die."[53]

I005 – Sunliner 403/F-18, USN

LCDR Michael Speicher of VFA-81, deployed aboard the USS *Saratoga,* launched as part of this strike.[54] Proceeding through Saudi airspace, his aircraft suffered a series of electrical malfunctions that degraded his ability to employ his HARMs. His radar warning system also failed, denying him the ability to detect enemy radars. Regardless, he pressed on with the mission.[55]

En route, Iraqi MiG-25s were reported airborne and active south of the formation near the airfield at Mudaysis. As Speicher's unit turned toward the MiG-25s, several pilots reported AAA near the airfield. A few minutes later a pilot in the strike package spotted an explosion and reported it as an air-to-air engagement. Two minutes later, the same pilot reported a larger ground explosion. He had not observed an ejection. Subsequently, Speicher was noted as missing when he failed to answer radio calls.

One of the pilots in Speicher's unit recalled the event. "I remember us all saying at that time, 'Where's [Speicher]?' . . . We checked a couple of times, tried different frequencies and got nothing. We called the AWACS to see if they had him. Nothing. I got on the radio with the JRCC in Riyadh via AWACS and told them we were missing an airplane."[56] The JRCC records do not show that such a call was received.[57]

Returning to the ship, one of the pilots immediately went to the intelligence center to report the loss. He stated that the explosion that he had witnessed did not appear to have been "survivable." Additionally, neither he nor anyone else had heard any calls on the emergency frequencies. But the pilots had been briefed that the Iraqis had the ability to "DF" or home in on survival radios, and it was possible that, if Speicher were down, he might have avoided using his radio for this reason.

As more aircrews recovered from the strike and nobody reported hearing from Speicher, hopes ebbed. Unbelievably, his loss was not immediately reported to the NRCC. Additionally, the coordinates finally reported were wrong.[58]

On this first strike, two aircraft were reported down. The second one was logged as incident **I006** and was supposedly an A-6. It took several hours of hard work by the SAR controllers to realize that the two reported incidents were, in fact, the same aircraft.[59]

***

Numerous crews had close calls. A flight of six F-111Fs was fragged to attack the Ali Al Salem Airfield in Kuwait. As they approached, they could see a wall of tracers surrounding the field. Heeding Glosson's words that "there's not a . . . thing in Iraq worth you dying for," three crews aborted and went to safer secondary targets. One crew who continued the attack had about all they could handle. As the pilot and WSO (weapons system officer) set up to drop their laser-guided bombs on the airfield, they were targeted by a dangerous SA-6 battery. Their bombs had just hit the target when the battery launched a missile at them. They took evasive action by diving to the deck. The missile passed overhead. The terrain-following radar on their F-111 had failed and they had no idea what their altitude was. Flying by instinct, the crew was beginning to believe that they were safe when an SA-3 missile battery locked onto them and fired. Again, they took evasive action and the missile streaked overhead. This made them feel invulnerable until they remembered that this was the first day of the war, and there were undoubtedly many more missions to come.[60]

Two more waves of F-117s and cruise missiles hit key targets before sunrise, and as the sun was coming up, several B-52s entered Saudi airspace. The aircraft had flown nonstop from Barksdale AFB, Louisiana. Entering preselected orbits, they began to launch conventional air-launched cruise missiles (CALCM). These unmanned miniature airplanes, each costing $380,000, scattered in all directions and destroyed six communications sites and electrical power plants across Iraq. It was their first use in combat. Completing the drop, the aircraft turned west and returned to home base. It was a 35-hour round-trip.[61]

The first night had gone well. LCDR Speicher was the only loss, reportedly at the hands of an Iraqi interceptor. He would turn out to be the only loss to an enemy aircraft in the war.[62]

Back at the SOCCENT operations center, Colonel Gray was elated. He had rejected Captain O'Boyle's "laager" idea to have helicopters and crews actually prepositioned in Iraq and still felt that, given the threat posed by the heat-seeking missiles, it was just too risky for the crews. He had agreed, though, to launching helicopters and having them orbit along the border while strike packages were in Iraq. They did not get a call on Sunliner 403. The predictions of heavy losses the first night had not been borne out.[63]

General Horner was also pleased. That morning, he started a ritual that would see him through the war. Each day when he came into the TACC, his first stop would be the JRCC corner of the center. He did this for several reasons:

1. He wanted to know what the losses were and what CSAR missions were laid on or how they were progressing.
2. He wanted to be reminded daily in the starkest of terms that war was about killing and that all of this was costing the lives of some of his men.
3. He wanted to keep faith with his aircrews because he knew how important the expectation of rescue was to the aircrews.
4. He wanted to know what was working and what was not so that he could stay ahead of the enemy in his targeting and overall strategy.[64]

Unaware of the mix-up over Speicher, Horner passed the results of the first strike to General Schwarzkopf. Being cautious, Horner had told him to expect up to 20 percent casualties. Instead, he called to report, "all aircraft back." Schwarzkopf thanked him.[65]

Out at the rescue alert sites, the rescue crews were perplexed. Cocked and ready to go, they were amazed that they were not being scrambled. Throughout the night several of them made radio checks with the SOCCENT command center to ensure that their radios were working properly. They were expecting the sky to "rain parachutes," as so many had predicted. As they watched the aircraft coming and going from Iraq, they were relieved that their services were not needed.[66]

A few hours later at a press conference in Washington, Secretary of Defense Cheney stated that only one aircraft had been downed on the first night and it involved a single casualty.

A reporter pressed him whether the casualty had been a wounding or death. Briefed on the incident by RADM Jeremy Boorda, the chief of naval personnel, Cheney replied, "A death."[67]

During the day, B-52s and Tornadoes pounded airfields in southern Iraq. The Tornadoes—flown by Great Britain, Saudi Arabia, and Italy—had a dangerous mission. They carried the JP.233, a dispenser munition specifically adapted for airfield attack. Each dispenser carried a number of shaped charges that deployed on parachutes. As they floated to the ground, a rocket charge would propel them into the runway. A delay fuse caused them to explode below the runway and heave up the concrete. This damage was time-consuming to repair.[68]

For proper employment, the weapons had to be released right over the airfield at 200 feet during straight-and-level flight. So far several attacks had been made, and no aircraft had been damaged, but the enemy reaction had been violent. One British pilot said, "I commented to my backseater on the heavy AAA in the two o'clock when we turned at point J where we changed from parallel track to twenty, second trail, and since *Jane's All The World's Fireworks Displays* was now in the twelve o'clock, it became apparent to both of us that the AAA was, in fact, coming from our target . . . deep joy!"[69] Their first loss would come soon enough.

1008 – Norwich 02/Tornado, RAF

The crew of Flt Lt John Peters and Flt Lt John Nichol was part of a four-ship flight attacking the Ar Rumaylah Airfield in southern Iraq.[70] It had been scheduled to hit this target during darkness, but a three-hour delay pushed them into daytime. During the intelligence briefing they were told that the airfield was lightly defended. In fact it was surrounded by 170 enemy AAA guns of various calibers and an unknown number of heat-seeking-missile sites. Carrying eight 1,000-pound bombs, they planned to make a loft attack from about 100 feet above the ground while doing 540 knots.

As they approached their pull-up point 25 miles from the target, they began to see enemy fire. The planned pull-up point would have given their bombs a full spread as they then fell on the airfield. The fuses were set to explode the bombs above the ground for maximum effect against the enemy guns. As they pulled up in their attack, though, their weapon release computer failed and did not release their bombs. They turned away from their target and jettisoned their ordnance. In the turn, the aircraft was struck by a suspected SA-16 heat-seeking missile. It hit the right tailpipe and destroyed that engine. Flames engulfed the right side of the aircraft. Flt Lt John Nichol called on the radio that they were on fire. The AWACS aircraft in that area monitored that call. Nichol then radioed that they were ejecting, but nobody heard that call.

Simultaneously, both men pulled their ejection handles. They landed close together and rapidly joined. Lieutenant Nichol took out his survival radio and made several radio calls; none was acknowledged. Their beacon was detected, but without voice contact, rescue forces were not dispatched.

There was no place to hide on the flat desert. They walked away from the crash site for close to an hour before about 30 enemy troops surrounded them and took them prisoner.[71]

Not long afterwards, Horner called General Schwarzkopf to update him on the air campaign. He told him of the loss of the Tornado and the earlier loss of Speicher. Schwarzkopf remarked, "My God, that is an order of magnitude better than what we expected."[72]

A-10s were also active, hitting remaining radar sites along the border. Other coalition forces were hitting targets in Kuwait. The French Jaguar units had received notification from their government just 15 hours before the start of the war that France would participate in the campaign. Hastily, they were given intelligence data so that they could participate in the first attack on Al Jaber Airfield in Kuwait. Trained in low-level tactics, they refused to coordinate with the American units hitting the field. Instead they went in at 100 feet and attacked just as preceding American flights were finishing. Every French aircraft was hit by ground fire, and one pilot was wounded.[73]

I009/M001 – Jupiter 01/Jaguar, FAF

The eastern sector AWACS detected an emergency radar squawk and a voice report of a bailout.[74] They reported it to the JRCC who requested the Marine RCC to launch two CH-46s.

Subsequently, AWACS determined that the aircraft involved were two French Air Force Jaguars, one of which had been hit by an SA-7 heat-seeking missile. They were okay and had landed at Al Ahsa Air Base, Saudi Arabia. The rescue effort was terminated.[75] All coalition air forces were now involved in offensive operations.

1010 – Bergan 23/A-4, Kuwaiti Air Force

A flight of four Kuwaiti aircraft attacked targets in Kuwait.[76] One aircraft was struck by AAA. The pilot ejected and was recovered by Kuwaiti resistance fighters. He remained with them and participated in several clandestine operations until they were all captured a few weeks later.

Proven Force

That afternoon the government of Turkey finally gave approval for the Proven Force task force to begin operations. As the crews scrambled to prepare for their missions, they were inundated with last-minute information. A team of survival instructors and pararescuemen quickly visited the units and gave updated briefings on CSAR techniques and procedures as the crews tried to absorb all of the detailed information necessary for the missions that they would fly that night.

The Proven Force combat rescue assets were also directed to deploy forward and prepare for CSAR operations. The MH-53s and MC-130s departed from Incirlik and flew to Batman. There they discovered that their first duty would be to literally set up their tents. Working quickly, they built a rudimentary command post. It only had one phone line, but they were able to establish SATCOM communications with SOCCENT. The crews then assumed combat recovery alert in support of SOCCENT and so notified them.[77]

With the forces now available in Turkey, Col George Gray felt comfortable that the coalition had a sufficient rescue capability throughout the area. He decided to split Iraq into two sections for rescue purposes. His forces in Saudi Arabia would handle combat recoveries up to 33°30' north latitude. Above that line the combat recovery forces with Proven Force would get the call.[78]

Photo courtesy of Steve Otto

One of the MH-53 pilots assigned to Proven Force, Capt Steve Otto, was involved in the rescue attempt of Corvette 03.

It was not quite that simple, though. Batman did not have secure access to the intelligence sources back at Incirlik. The crews had brought some data on the Iraqi threats with them. They quickly began planning possible operations into the northern sector. One of the MH-53 pilots, Capt Steve Otto, was involved in the process and recalls:

> We figured out early on that the threats and the terrain of northern Iraq were much different than southern Iraq. Down south, we anticipated that we would be able to drive around the Iraqi threats, to use our refueling and range capability, if not to defeat detection, then certainly to defeat the threat systems. Up north we realized that we were not going to have that luxury, that there were only 13 mountain valleys that went from Turkey into Iraq. The threats were pretty robust up there, and we were going to have to fly at night. . . . We knew from the threat that there was a great concentration of Iraqi threats, a lot of ground divisions, AAA, and SAMs down in the tri-border area where Iraq, Turkey, and Syria meet. We knew that because of threat considerations, we would have to fly far east if we were going to fly through Iraq.[79]

Regardless, the men at Batman were now ready. But as another of the MH-53 pilots, Capt Grant Harden, recalled, they intended to make their flights into Iraq at night for obvious tactical reasons.[80]

That afternoon a strike force of 32 F-16s hit the Al Taqaddum Airfield and a nearby petroleum refinery. They were supported by 15 F-15s, eight F-4Gs, and four EF-111s. This was followed by massive B-52 strikes on Republican Guard divisions in Kuwait. As darkness set in, F-117s attacked air defense sites in western Iraq. Unfortunately, bad weather had moved in, and several targets were not damaged at all.

Starting at 2200L, a series of raids hit targets in the Basra area just north of Kuwait. The strikes were heavily supported with jamming and SAM suppression. There were still some losses.

I011 – Norwich 21/Tornado, RAF

This was a strike by four Tornadoes against the Shaibah Airfield in southern Iraq using the JP.233 cluster weapon specifically designed to destroy runways.[81] The crew of Wg Cdr Nigel Elsdon and Flt Lt Robert Colier had apparently made a successful run on the airfield but was engaged by a SAM site upon egressing the target. Another crew in the flight observed an explosion and asked its controller on AWACS how many radar returns he was tracking. The controller replied that he had only three. No contact was ever made with the crew, and intelligence later determined that both men had been killed. Elsdon was the most senior coalition officer killed in the air campaign.[82]

It was becoming rapidly obvious to the operations planners that southern Iraq was a dangerous place. Regardless, the targets had to be hit.

I012 – T-Bird 56/F-15E, USAF

A flight of six F-15Es attacked another airfield west of Basra in southern Iraq.[83] Each aircraft made a run over the target in sequence. The number three aircraft (T-Bird 53) had to abort his run-in because of heavy enemy AAA. Coming around for another attack, the pilot made a radio call to inform the crew of the sixth aircraft that they would not be the last ones off the target. Lining up for his attack the pilot heard the crew of T-Bird 56 call that they were clear of the target. As the pilot of T-Bird 53 pressed his attack, he observed a large explosion. No Mayday call was heard. On egress from the target, T-Bird 56 did not check in. This was reported to AWACS with a general location of where the explosion occurred. AWACS notified the JRCC. With no indication that the crew was alive, no rescue effort was mounted, but the JRCC directed search operations in the area for several days. Contact was never made with the crew of Maj Tom Koritz and Maj Donnie Holland. In fact they had been killed in the engagement. Koritz was dual-rated as a pilot and flight surgeon.[84]

Strikes were also being flown in western Iraq, primarily by US Navy units in the Red Sea. Most were large packages that included aircraft for MiG protection, general antiradar jamming, and SAM suppression. They would precede the attack aircraft, which would then deliver bombs against the actual targets. This tactic was not always completely successful.

I013 – Quicksand 12/two A-6s, USN

These two aircraft from VA-35 launched from the USS *Saratoga* and were part of a large strike package hitting the heavily defended H-3 airfield in western Iraq.[85] Leading the force were F-14s for air-to-air protection and F-18s and EA-6Bs to attack and suppress SA-2, SA-3, and SA-6 missile sites in the area.

The A-6s made low-level attacks at 400 feet from different attack axes. One of the aircraft, crewed by LT Rob Wetzel and LT Jeff Zaun, set up to deliver a load of Mk-20 "Rockeye" on the POL (petroleum, oils, and lubricants) storage facilities near the airfield. The Iraqis had been alerted and flooded the skies over the airfield with parachute flares, creating a bizarre and disorienting scene. The pilot, Wetzel, described the run-in: "We turned at the IP [initial point] 30 miles away at 450 knots, flying at 500 feet. It was almost a daytime scenario. SAMs flew around the strike force. There was heavy AAA. I saw a missile go off in front of us. I saw a missile to the right at two o'clock, turned into it and dropped chaff. We were hit."

The aircraft was hit by either an SA-6 or Roland missile and crashed north of the airfield. Wetzel continued, "The back of the airplane was on fire. We ejected through the canopy at 500 feet and 400 to 500 knots. Our helmets were ripped off. We had flail injuries."[86]

To Zaun it was all just a blur. Neither he nor his pilot was able to make an emergency radio call before the aircraft began to come apart. He said that,

> [The missile] threw us to one side. Made a loud sound. It hit my side. The right engine ate itself. I looked to see how high we were. There was a lot . . . going on. When I punched, I was sure it was the right thing to do. I remember looking at the altimeter. I could not figure it out. So I ejected just as they taught us. . . . I do not remember being in my parachute. I remember being in the sand and going "Whoa, ... It's war and I am in the desert." . . . I actually did calm myself down. I got up and patted myself down for injuries. The first thing I looked for was my radio. I had lost it in the ejection. That was a bad thing. That demoralized me as much as anything else. . . . I carried my parachute for some time because I thought the radio was in it. I heard my pilot calling me. He was hurt. He took off his LPA [survival vest] and did not have a radio either. I asked him where it was. . . . He was hurt, and in taking off his LPA had lost it [PRC-112 radio]. I was not happy. . . . I did not have a radio and could not tell anybody I was alive.[87]

Lieutenant Wetzel had a badly wounded shoulder. Realizing that there were enemy troops in the area, the two men started to evade to the west, away from the airfield. Zaun intended to find an open area and lay out some kind of signal that could hopefully be picked up by imagery intelligence.[88]

A USAF AWACS was in the area but did not establish any contact with the crew. After a short period of time, estimated to be no more than an hour, they were found and captured by a team of Iraqi soldiers who took them to the dispensary on the airfield.[89] Saudi intelligence subsequently reported that the crew had been captured.

The second A-6 was also heavily damaged. It managed to divert to Al Jouf Airfield in western Saudi Arabia. There, analysts determined that it had, in fact, been hit by a Roland missile. The crew was safe and uninjured, but the aircraft never flew again.[90]

This night had not gone as well as the first. Three aircraft were lost to either SAMs or AAA. All three crews had been killed or captured. Thirteen other aircraft had been damaged but managed to return to friendly control.

18 January

In place and operational, the aircrews of Proven Force, in Turkey, entered the campaign with strikes against strategic sites in the north. There were still some diplomatic problems, and one package of F-15s was not granted border-crossing authority by Turkish air traffic control. Of the aircraft that did reach their targets, one of the F-111Es was damaged by AAA.[91]

In the south, air strikes continued at the previous day's pace. Coalition aircraft attacked Iraqi forces in Kuwait and reattacked airfields that showed signs of activity.

1015 – Caesar 44/Tornado, Italian Air Force

Eight Italian Tornadoes took off to attack another airfield in southern Iraq.[92] The flight had problems with refueling, and only one aircraft was able to get gas. That crew pressed on to the target. During their bomb run they were shot down and immediately captured. One of the crewmembers literally landed next to a truck full of Iraqi troops. There was no beacon or voice transmission, and no parachutes were spotted. Upon notification, the JRCC scrambled to collect what little information it could through the liaison channels to the Italian Air Force. Since the loss was deep in enemy territory, the mission was passed to SOCCENT. Colonel Gray looked at it and immediately realized that the area was extremely well defended. He did not plan a rescue operation. When their POW status was confirmed, no further search efforts were made.[93]

***

During the day, the war took a curious turn. Unable to staunch the onslaught coming at his country, Saddam ordered his forces to launch Scud missiles at Israel. US satellites immediately picked up the launches, and this data was shared with the Israeli government. Fearing that the Iraqis would buttress this attack with an air strike, the Israelis began preparing to launch their F-16s. Only direct intervention from President Bush kept the Israeli aircraft on the ground.

That day, seven missiles landed in Israel. They did relatively little damage, but the political impact was considerable. General Schwarzkopf started getting calls from Washington to do something to stop the Scuds. In fact F-15Es and several packages of naval aircraft had already hit the launch sites in western Iraq. It was now becoming clear, though, that perhaps more—much more—would have to be done about the missiles.[94] The strategic air campaign plan, so skillfully crafted and molded, was not surviving contact with the enemy.

Coalition aircraft had also begun bombing Iraqi units in Kuwait. The US Marines launched OV-10 forward air controller aircraft to look for targets. The Iraqi gunners spotted them and waited for their chance to act.

I016/M002 – Hostage 75/OV-10, USMC

Lt Col Cliff Acree, squadron commander of Marine Observation Squadron Two (VMO-2), and CW04 Guy Hunter were shot down 14 miles northeast of Mishab, Kuwait, while searching for targets.[95] They were flying at 8,000 feet and looking for Iraqi artillery and FROG (free rocket over ground) sites when their aircraft was hit by a heat-seeking missile. The pilot noticed the missile just as it hit the exhaust stack on the right engine. The explosion destroyed the engine and ripped the wing from the fuselage. He quickly made a Mayday call; then the aircraft began to come apart, and both men successfully ejected. Descending, the pilot could see that they were floating down right over Iraqi ground units. Landing, he ran to his backseater who was slightly wounded. Acree had to make a decision as to what to do.[96] He described his thoughts,

> We'd been trained to escape and evade if downed in enemy territory. But our chances of evasion were slim to none. We couldn't have picked a worse spot to land if we'd tried—we were in the heart of an Iraqi division that had taken a serious pounding by Allied forces the previous day, and were within a mile of the artillery batteries we'd been trying to trash ten minutes before. The area we'd landed in was so heavily fortified that you could run half a mile in any direction and bump into somebody or something. . . . Our only hope was to get picked up by a search-and-rescue ([C]SAR) helicopter. We pulled our small survival radios from our torso harnesses. Mayday! Mayday! Mayday! We tried to get *anybody* to respond. No response, only static. The Iraqis would soon arrive—time was running out. I scanned the sky. A Huey helicopter with two Cobra escorts would be a gift from heaven right now.[97]

Instead of friendly helicopters, Acree saw enemy vehicles and troops closing in on them. Within minutes they were surrounded. Acree kept calling on his radio, "Downed aircraft. . . . Acree and Hunter are alive and on Kuwaiti soil." The soldiers slowly closed in around them. Acree considered trying to run, but with a wounded comrade, he realized that it would be futile. The reality of the situation was starkly clear. "There's a time when you have to call it quits. Staring down the barrels of those AK-47s, I knew this was it."[98] The men surrendered and became POWs. They had been on the ground less than 10 minutes. Other Iraqi forces swarmed into the area and began setting up defensive positions to fight off any rescue forces.

When notified, the JRCC tasked the mission to the NRCC on the USS *Blue Ridge*. The NRCC launched helicopters from the USS *Leftwich* to execute a search, even though neither an emergency beacon nor a radio call from the crew had been heard.[99] The search was joined by an MH-3 from the 71st SOS, a RSAF helicopter, two Saudi patrol boats, and three Saudi fast boats. Later that night a US Navy P-3, a Royal Navy Nimrod, and a US Navy E-2 continued to search. With no positive results, the search was abandoned the next day.[100]

Losing its leader was a terrible blow to VMO-2.[101] OV-10s flown by forward air controllers had the job of directly coordinating with the ground forces and were not supposed to venture north of the Saudi-Kuwait border unless it was an emergency situation. What constituted an "emergency" was not clear. The downing of Hostage 75 tragically illuminated the reason for that restriction and made the loss difficult to justify.[102]

I019 – Iguana 70/MH-3, USAF

An MH-3 from the 71st SOS had an in-flight emergency but was able to recover at a friendly field. It had been performing a SOCCENT mission.[103]

I020 – Jackal 11/A-6, USN

In an effort to bottle up Iraqi naval units in the northern Gulf area, a large force of aircraft from Battle Force Zulu launched to drop aerial-delivered mines.[104] This was the first such operation since the Vietnam War. Four A-6s from VA-155 on the USS *Ranger*, with support from 14 other aircraft, made a low-level drop of 42 MK 36 destructor mines at the mouth of the Az Zubayr River. Jackal 11 was fragged, as the second of four A-6s, to make a low-level attack to deliver mines against the Umm Qasr Naval Base just north of Bubiyan Island on the Kuwait-

Iraq border. En route weather was poor, with low clouds and low visibility. Other aircraft in the formation observed AAA but no SAMs. Jackal 11 was observed maneuvering off of the ingress path when it went down. There was no emergency call from the crew. A few minutes later an orbiting RC-135 Rivet Joint aircraft reported an emergency beacon in the vicinity of Bubiyan. A US Navy E-2 Hawkeye also in the area reported voice contact on one of the emergency frequencies. The voice was accented and believed to be an attempt to lure rescue forces into a trap.[105]

While all of that was being sorted out, other aircraft from the *Ranger* formed an ad hoc task force and rendezvoused with a Navy helicopter for an immediate pickup attempt. Several F-14s and A-6s took up an escort formation over the helicopter as it approached the coast. The swarm of aircraft was greeted with heavy AAA fire, and somebody in the formation called, "Abort, Abort!" The helicopter did a 180-degree turn, and all of the aircraft went back out over the water. One of the A-6 pilots was befuddled at the whole maneuver and later wrote,

> We were all left wondering what . . . had happened. Had a survivor actually been authenticated via radio call sign, and if so, did we have a good geographic cut on where he was? Without a clear picture of the survivor's location, it was suicide to wander low over the open skies of Iraq looking for him and dueling with every AAA site along the Gulf. Back in the debrief, the intel[igence] guys were not much help in filling in the blanks. It seemed that no one could say with certainty whether or not a survivor had been authenticated as legitimate, yet we had been prepared to risk five aircraft on a rescue mission to a phantom man on the ground.[106]

SOCCENT also initiated mission planning, but with no valid voice contact or position, and considering the enemy threat in the area, no follow-on rescue effort was launched. Search operations continued in the area for several days. Subsequently, intelligence sources determined that the crew of LT William Costen and LT Charles Turner had been killed.[107]

Intelligence studied the intrusion and determined that enemy forces had gotten to the wreckage and were trying to lure rescue forces into the area to shoot them down. Planners at CENTCOM decided to develop a plan to set up a fake CSAR and lure some Iraqis themselves into a trap. Such a plan had to be vetted through the CENTCOM staff. The plan sounded feasible, but the staff lawyers objected because they felt that it violated the international principle that emergency signals should not be used for tactical purposes. The operation was never run.[108]

Studying the results of this strike, Navy intelligence determined that the Iraqi navy was already bottled up because of mines that it had laid itself. After realizing how dense enemy air defenses were in the area, VADM Stanley Arthur, commander US Naval Forces Central Command, decided to cancel any further mine-laying flights.[109]

Admiral Arthur also had his staff look at the efficacy of low-level operations overall in the first two days of attacks. The two A-6s shot down had been lost on low-level attacks. Addressing this and the losses of the Tornadoes and the F-15E—all on low-level operations—he stated to his commanders:

> Gentlemen, far be it from me to dictate specific combat tactics. But I must inject my early observations relative to the age-old argument of low-altitude delivery versus high. With a quick look at what has happened to the multinational forces to date, one cannot escape the fact that current AAA environment makes the low-level delivery a non-starter. I want you to take a hard look at how your air wings are handling the issue. We learned a hard lesson in Vietnam relative to AAA and then later many told us we learned it wrong—I think not. There is a place and time for low-altitude delivery and it usually involves surprise. We can no longer count on surprise. That went away shortly after [midnight on D-day].[110]

The air wing commanders from the carriers directed their aircrews to switch to medium-altitude operations. This had a dramatic impact on mission effectiveness because the crews had not trained to deliver ordnance from higher altitudes. Additionally, the carriers only had limited numbers of laser-guided bombs, which were the most effective weapons for medium-altitude operations. What laser-guided bombs were available were to be carried by the A-6s. The Navy had not brought any Maverick guided missiles that could also be effectively used at the higher altitudes. Consequently, the only ordnance available for the F/A-18s were the old Vietnam-era "dumb" bombs that were much less accurate when dropped from higher altitudes.[111]

I021 – Falstaff 66/F-4G, reported as a lost communications, USAF

The Iraqi gunners were not the only threat to the coalition aircraft. In the winter the Persian Gulf nations are occasionally covered with dense fog. The cover can be so thick that the pilots cannot see the runways. This was another worry for the aircrews. In actuality this aircraft and crew from the 52d TFW crashed due to fuel starvation.[112] The crew had tried to land at several airfields that were below landing minimums because of dense fog. The crewmen ejected safely and were recovered by base personnel. A subsequent investigation determined that enemy fire had ruptured a fuel tank on the aircraft, causing the fuel starvation. The aircraft was subsequently listed as a combat loss.[113]

***

Taking a note from the Navy, Air Force planners did a quick analysis to determine the validity of flying at higher altitudes. Intelligence data indicated that the combination of jamming, suppression, precision strikes on command centers, and constant attacks on the Iraqi Air Force had dramatically reduced the threat above 10,000 feet,[114] and sufficient stocks of laser-guided bombs and Maverick missiles were available to ensure target destruction.

General Horner left it up to the individual wing commanders to decide what tactics to use. The B-52s, after having some of their aircraft damaged by enemy AAA and SAMs, were ordered by their commander to change to high-level tactics. The Tornado units refused to change. They had to make low-level attacks to deliver the JP.233 weapon most effectively against runways.[115] Overall, the campaign seemed to be going well.

Aboard the *Saratoga*, Lieutenant Scudder was growing concerned about the PRC-112 radios. When the shipment had arrived, he had noted several corroded batteries and had not received many spare batteries. Now some of the crewmembers were reporting that the radios had a tendency to slip out of the modified pouches on their survival vests. Some pilots had noted that the radio was too easy to turn on. That and the constant testing and re-keying with personal codes was wearing out the batteries. Several pilots asked for extra batteries. None were available.[116]

19 January

Weather became a major factor on day three of the campaign. Overnight several F-117 strikes could not hit their targets because of clouds. During the day numerous large packages of aircraft were scheduled to hit key targets in the Baghdad area and around Basra, but the air tasking system was experiencing growing pains and could not stay up with the constant changes being forced by weather, mechanical delays, and enemy reactions. In the morning, one large package of B-52s and F-15Es hit two Republican Guard divisions in the Kuwaiti theater of operations (KTO). In studying the results of the B-52 strikes, the intelligence specialists were beginning to pick up a trend. In several instances the B-52 strikes were 400–600 feet short of their aim points. Prior to the war the B-52s had been given an upgraded inertial navigation system (INS) for all-weather delivery and, theoretically at least, their bombing accuracy should have been higher. The analysts were puzzled until one discovered the problem. The targeting cell in the TACC had passed target coordinates based on the GPS WGS84 datum, but the aircraft INS had been initialized on the ramp at Diego Garcia, where the coordinates given to the crews had been based on the WGS72 datum. The difference between the two exactly matched the miss distances for the bombs, and the error was quickly corrected.[117]

In midday a large Navy package hit petroleum sites west of Baghdad. A force of 40 F-16s with a large support package of EF-111s and F-4Gs was scheduled behind them and directed against the military intelligence headquarters in the outskirts of the capital. That package was diverted to the west to strike Scud sites near the H-3 airfield as part of the reaction to that threat.[118]

Package Q

Another force of 72 F-16s from the 388th TFW and 401st TFW was fragged to hit a nuclear research facility southeast of Baghdad. Overnight the target was changed to a series of three Iraqi government ministry buildings. All were large targets that needed to be hit by many bombs. They were widely scattered, and the attack was organized so that they were struck sequentially starting with the one farthest from the heart of the city and working inwards.

In planning the missions, some of the unit commanders compared these to the mass raids against heavily defended Hanoi in the war in Southeast Asia. The raw numbers of SAM sites and AAA guns were alarming, but the Baghdad area defenses had been so heavily pummeled in the last two days that an air of cockiness was working its way through some of the units and fighter crews.[119]

The geographical arrangement of the new targets meant that the enemy defenses would be fully alerted as the strikes progressed. Perhaps more critical, the mission change had not reached the crews until they were almost ready to go, so the flight leads could not make any fundamental changes to their tactical plan prior to launch.

To some, the last-minute changes introduced a healthy dose of risk to the mission. The package included several F-4Gs for SAM suppression. That heartened the strike pilots because they knew how effective the Weasels were in suppressing the sites with their HARMs. Because of the distance to the target, each F-4G carried only two HARMs instead of the usual four, reducing their capability by half. The flight leads were unaware of this critical fact.

Once airborne all aircraft had to be in-flight refueled. Here, too, there were problems. There was rough weather in the refueling track. Some of the tankers had to throttle back because of winds so that they could refuel all the aircraft prior to the end point of the track. Things got so fouled up that some flights were unable to refuel and had to return home with their bombs. Those that were able to refuel pressed on north and broke out into open skies over Iraq.

I023/M003 – Stroke 65/F-16, USAF

Maj Jeff Tice of the 614th TFS was leading a flight of four F-16s in the last wave of this 72-plane attack force sent to hit targets in the Baghdad area.[120] The flight was preceded by eight F-15s for air superiority and eight F-4G Wild Weasels for SAM suppression.

Weather caused delays en route, and by the time the F-16s were approaching their targets, the F-4Gs had already fired all of their SAM-suppression missiles and were departing the area.

Photo courtesy of Paul Harmon

Crewmembers on the Stroke 65 recovery attempt, shown here on 22 January 1991, include: *back row, left to right,* Sgt Rob Turner, Capt Paul Harmon, Capt Ken Sipperly, Sgt Jeff Mucha, SSgt Mike Hulin, and TSgt Troy Arce and, *front row, left to right,* Sgt Ismael Gonzalez and SRA Martin Brown.

As the Weasels were leaving, numerous SAM sites went active and began to track the aircraft. The flights began to maneuver individually in self-defense. Various missile batteries began to fire at individual aircraft, and each pilot maneuvered to avoid the deadly weapons, causing the flights to become badly scattered. Major Tice was fired on by an SA-3 missile battery. Seeing the missile, he maneuvered to evade it. The missile shot by and detonated. Unfortunately, Tice's aircraft was within the lethal range of the warhead and was fatally damaged. Quickly scanning his aircraft, he could see fuel, oil, and hydraulic fluid seeping from numerous holes.

As Tice turned his aircraft to head south, warning lights were flashing in his cockpit, and flight instruments were beginning to fail. His wingman, Capt Bruce Cox, joined on his wing to escort him out of the area just as AWACS called and informed them that Iraqi MiG-29s were approaching from the north. Cox did a hard right turn and picked up the enemy aircraft. He quickly targeted one of them with his radar. The MiG spotted him and turned away. Low on gas, Cox could not pursue. Instead he turned back south and tried to rejoin with Major Tice.

Moments later Tice's engine failed. He stayed with his aircraft as long as possible before ejecting. He was 150 miles inside Iraq.

As he descended, he grabbed his emergency radio and reported that he could see enemy personnel watching from below. He was difficult to hear because the enemy was jamming the rescue frequency. Tice asked for aid from his flight, but all were extremely low on fuel and could not remain. He was captured by Bedouin tribesmen as he landed.[121]

Unaware of his fate, the flight passed his location to AWACS. They reported it to the JRCC, who tasked the mission to SOCCENT. SOCCENT analyzed the situation and determined that a rescue in the area was possible. As diverted F-15s orbited Tice's reported position, an MC-130 searched the area but did not make contact with Tice.

Fortuitously, about 170 miles south, two MH-53s of the 20th SOS were flying from ArAr to Rafha. The flight leader was Capt Tim Minish, call sign Moccasin 05. The second aircraft was flown by Capt Paul Harmon. En route, they were contacted on their tactical radio by Colonel Comer. Comer gave them initial rescue data and ordered them to quickly refuel the two aircraft at Rafha.

They landed and took their aircraft through the refueling pits. Based upon the reported position of the survivor, Minish planned the best ingress using a preplanned "spider route" which he personally had helped develop with CENTAF intelligence.

When both aircraft were fueled, they headed north. Each had a crew of eight, which included two Air Force PJs. As they approached the border, the weather began to deteriorate, with decreasing visibility and lowering ceilings. In the weather, Captain Harmon was flying one-half rotor width from the lead aircraft at about 80 feet above the ground. They entered an especially thick fog bank, and Harmon lost sight of the lead aircraft. Automatically, he moved away from his leader's last known position to ensure separation and transitioned to flying by his own instruments. He was then able to acquire the lead aircraft on his radar and fell into a trail formation.[122]

Two MH-60s from the 3/160 SOAR also launched. They proceeded to an orbit point near the border in case Moccasin and Harmon needed any help.

Approximately 40 miles from the survivor, Moccasin 05 contacted the F-15 CAP. The flight lead (Kayo 01) was the on-scene commander and gave them a quick area briefing. He told them about an active AAA site five miles to the east. Minish asked AWACS for a strike flight to hit the AAA site. A flight of F-16s checked in with Minish, and he gave them a rough location for the enemy guns. The F-16s tried to get below the weather to find the guns but were unable to do so and left the area.

Ten miles from the survivor's location, the crew of Moccasin 05 started calling him on the emergency frequencies. Above, Kayo 01 did the same. When Stroke 65 did not respond, the helicopter crews started a sector search. That proved fruitless though they searched the area for about 30 minutes and flew over several Bedouin encampments. They reported this to AWACS, and the controller gave them another set of coordinates farther south where aircraft wreckage had been sighted.

They headed to the new site, calling for Stroke 65 as they went. He never responded. Approaching the new location, "Shawnee" called and informed them that the Iraqi sector defenses had detected their presence through the use of radio direction finding. Neither Minish nor Harmon knew who Shawnee was, but after spending so much time deep in enemy territory, neither was inclined to question the call. With AWACS closely monitoring their flight path, the two MH-53s turned south and departed. The weather was still terrible, and again, they had to use their terrain-following radar to allow them to stay low.[123]

Crossing the border they were briefly illuminated by a spotlight and observed some AAA well behind them. They landed at Rafha uneventfully just ahead of a slow-moving fog bank. The crews had logged more than five hours in enemy territory.[124]

All efforts to contact Tice were futile because, upon capture, he was stripped of his emergency gear and radio. As the aircraft searched, he was a prisoner held captive in a Bedouin tent. He could only listen as the aircraft passed overhead.[125]

While the effort to rescue Major Tice had been unsuccessful, the mission did mark a milestone. Captain Harmon wrote

> This was the first time that the Pave Low had been fully used as designed—Vietnam lessons learned, adverse weather, night capability, etc. It was actually the first time the airplane had ever done what it was designed to do. We never considered turning around because of the weather. We just continued north. Tim [Minish] had already configured his airplane to set up his hover coupler to descend through the weather had he needed to once we got into the area, had Tice come up on the radio.[126]

Colonel Comer was also impressed with the ability of the MH-53 to operate through the low clouds in enemy territory with relative impunity. The ability of the Iraqi defenses to effectively DF the survivors' radios and even partly track aircraft through their radio communications concerned him, however. That was a dangerous development and portended great danger for his helicopter crews. He pointed out that,

> You have a helicopter with six or eight people—which is a tasty morsel all its own—taking the risk to pick up the single person who might be out there . . . or two people. It is not a rescue, it's a war. . . . When you say rescue you think about a Coast Guard guy who went out there and picked up some guy who is in trouble. That is a worthy and worthwhile mission. But doing it while there is somebody out there trying to shoot you—that is a war. And that is what we were thinking. How do we deal with the war? How do we deal with the desert? How do we deal with all of these things that seem to be threatening to us while at the same time, we have to have the communication and wherewithal to know where a person is? Now we did launch and we did end up searching.[127]

The effort for Major Tice had been valiant if unsuccessful. Unfortunately, he was not the only pilot lost in Force Package Q.

1024 – Clap 74/F-16, USAF

Capt Harry Roberts of the 614th TFS was flying in the strike package with Major Tice.[128] In the same melee, Roberts' flight was engaged by an SA-2 battery but was able to evade its missiles. Weather had moved over their target, and they were not able to drop their bombs. Instead they turned south to egress and were engaged by a newer and much more lethal SA-6. Roberts' F-16 was specifically targeted by the enemy site. His radar-warning receiver lit up, telling him that he was locked in by the tracking radar. He maneuvered his aircraft and jettisoned his bombs and external fuel tanks in an effort to evade the threat. He was unsuccessful, and the battery launched two missiles at him. One missile guided true and detonated near his aircraft. He felt a little bump and saw smoke coming from below his left wing. His engine was still running normally. He tried to head south toward Saudi Arabia. Then his flight controls locked up, and he ejected at about 22,000 feet.

He free-fell until his parachute deployed, as programmed, at 14,000 feet. He was immediately struck by the relative silence. Above he could hear the jets in the distance, departing to the south. It was a lonely realization. Below, he could hear the occasional sound of the AAA guns still firing. Looking down, he could see tracer rounds streak by and a large four-lane highway running off to the south. He could see vehicles stopping along it and people getting out to watch him float to the ground. Not too far north was an occupied AAA site. He could see the gunners watching him. As he hit the ground in an open field, he grabbed his emergency bailout kit and tried to move away from the road and soldiers. The combined enemy group was able to cut off any escape route, and sensing his fate, he put up his hands.[129] Unfortunately, none of his flight mates saw him eject, and because of his quick capture, he did not get out an emergency call.[130] AWACS reported the loss to the JRCC, and they tasked it to SOCCENT, but without any contact with the pilot or any real knowledge of his location, no rescue effort was mounted.

Immediate tactical lessons were drawn from the loss of these two aircraft:

1. Large packages of aircraft could not be sent into the skies over Baghdad.
2. Bad things happened when aircraft were in SAM rings and the Wild Weasels were not available.
3. F-16 packages should be smaller in size so that they were more manageable.[131]

From then on, only F-117s and cruise missiles would hit targets in Baghdad.

I026/M004 – Dark 13/Tornado, RSAF

This aircraft had a communications problem and ran out of fuel as it tried to land at two airfields that were both below landing minimums because of fog. The crew ejected near Hafr Al Batin and was rescued by Saudi ground forces.[132]

Not all losses in war are to the enemy. All of the other factors—bad weather, mechanical problems, human failure—still apply.

I027 – Newport 15B/Tornado, RAF

Eight Tornadoes were fragged to attack the main airfield at Al Tallil near Basra.[133] This was a very heavily defended position with multiple AAA sites and a Roland missile battery. The Tornado crews decided to modify their tactics. The first four aircraft would carry 1,000-pound bombs that they intended to "loft" onto the airfield to suppress the defenses while the second four aircraft attacked the runway with their JP.233 cluster munitions.

The aircraft successfully rendezvoused with their tanker to top off their gas, then made a low-level run into southern Iraq. The approach to the target was deceptively quiet. As the first four began their "pop-up" maneuver to release their bombs, the airfield defenses discovered them and began to react. Immediately, the sky was filled with tracer rounds. Flt Lt David Waddington and Flt Lt Robert Stewart detected a Roland site directly in front of them. They saw it launch a missile at them. The pilot banked the aircraft sharply to the left and dropped chaff. The missile went off below the aircraft in a tremendous white flash. The blast shattered the canopy and knocked the pilot unconscious. Stewart, in the back seat, used the "command" ejection system to eject both of them. Stewart had a broken shoulder, a crushed vertebra, and a broken leg.

The other crews in the flight saw the aircraft go down and reported the loss and lack of contact with the crew. They were convinced that the crewmembers had been killed. In fact, both were still alive. They tried to signal with their survival radios, using both voice and beacon, but were never heard. After an estimated 14 hours, they were captured. Both were released after the war.[134]

In the TACC, General Glosson felt that perhaps the Brits were on the ground and evading. He wanted to send in recovery forces and discussed it with Capt Randy O'Boyle. O'Boyle argued that the area was far too dangerous for any kind of helicopter search, especially since they had no contact with either crewmember.¹³⁵

The loss of another Tornado sent shock waves through the air planners. At the rate they were being lost, the Tornadoes would not survive the war.

***

The euphoria that the commanders had felt after the first two days of the air war quickly evaporated. The steady loss of the Tornadoes and the double loss of Stroke 65 and Clap 74 made the planners realize that they did not yet have the key to victory in this conflict.

Now another dilemma was developing. Apparently Saddam had decided to step up the use of his Scud missiles. During the 1980s the Soviet Union had sold Iraq large numbers of these short-range missiles. They had fired 190 of them against Iran in their border war. The missiles had a range of 600 km, with a 180-kilogram warhead and a CEP (circular error probable) of 2,000 meters. Tactically speaking, they were not worth much.¹³⁶

But that was not the problem. With that range, they could hit any country in the Gulf region—including Israel. If Israel chose to respond, the Scuds would become a political and hence strategic problem, because an Israeli response could affect the delicate coalition so carefully crafted to defeat Saddam and Iraq.

Since the first Scud launches 18 January, Horner had directed that air strikes be sent to attack the launch areas in western Iraq. With each missile launch, more flights were diverted from the Baghdad area to the vicinity of the H-2 and H-3 airfields. But the launches did not abate, and the Israelis had begun to react. At one point they had launched a large force of F-16s to hit the sites in western Iraq. Frantic calls from Washington convinced Israel to recall the fighters. General Schwarzkopf became aware of Israeli plans to launch a sustained campaign against the sites, which would even include attack helicopters, and all of this would proceed through Saudi airspace.¹³⁷ He ordered his airmen to "do something."

General Horner swung into action and began to modify the air campaign plan. The following sequence was recorded in the TACC log on 18 January:

> 0825 – General Glosson on the phone in the Black Hole: We will spend the remainder of the day targeting Scud sites.
>
> 0938 – General Glosson: CINC [Schwarzkopf] is getting a lot of calls from Washington about the Scuds.
>
> 0948 – Second launch of A-10s found seven TELS [mobile launchers]; destroyed two. We are sending more A-10s.
>
> 1040 – [Col] Crigger to Horner: A-10s are being sent to seven TELS. Also the F-15s are on the way. . . . Horner: They (F-15s) should be there by now. Doesn't care if they get there all at the same time. Want those Scuds gone.[138]

Glosson immediately directed the diversion of more flights into the western areas. A contingent of A-10s was deployed west to Al Jouf to maintain a constant presence over that area. Given the physical nature of the missiles and their fixed- and mobile-launch systems, planners determined that the best weapons to send against the Scuds were the F-15Es of the 4th Wing at Al Kharj Air Base. The air planners began diverting them wholesale to the Scud targets. Like the day before, over Baghdad, the flight leaders did not have a lot of time to fully plan and brief their missions.

1028/M007 – Corvette 03/F-15E, USAF

A package of 24 F-15Es was diverted from other preplanned missions to hit Scud missile sites in western Iraq.[139] Coming less than six hours before takeoff, the changes caused near chaos among the crews as they scrambled to get proper intelligence and request support assets to attack the sites. The area was extremely dangerous because of all the enemy defenses there. The aircrews nicknamed it "Sam's Town," the name of a nightclub in Las Vegas, for the preponderance of SAM sites.

Flying as the number three aircraft in Corvette flight was the wing director of operations, Col David Eberly, and Maj Tom Griffith, who was considered one of the best WSOs in the unit.

Only as the flight crews walked to their jets were they finally handed the actual aim points to be used for their bomb deliveries, and the necessary times over target (TOT) were changed

Photo courtesy of Dave Eberly

Col Dave Eberly was the ill-fated pilot of Corvette 03.

Source: Darrel Whitcomb

The back-seater in Corvette 03 was Maj Tom Griffith, shown here in a 2005 photograph.

twice. The flight leader, Lt Col "Scottie" Scott remarked to Colonel Eberly, "This thing is a goat rope. It's the kind of mission that gets people killed."

Silently, Eberly concurred but understood the importance of the mission. He knew that there were SA-2s and SA-3s in the target area, and the ALQ-135 jamming pod on his aircraft did not have jamming capability against those weapons. Corvette flight would have to rely on the supporting F-4Gs and EF-111s to suppress the SAM sites.[140]

After takeoff, the situation progressively worsened. First they had trouble finding their refueling tankers in the thick clouds. Then they found out that the F-4G Wild Weasels scheduled to support them had not been given the new TOTs and would not be with them as they entered the target area.

As the strike aircraft flew west, they were again hit with misfortune. Colonel Scott had requested jamming support from EF-111s. A flight of two had, in fact, arrived on station and set up their orbits to electronically jam the Iraqi missile sites. But an Iraqi MiG-25 took off, intent on downing one of them. As it darted through the two EF-111s, it fired three missiles. Both pilots took evasive action and the missiles missed. The maneuvers drove the two jammers out of their orbit. Not knowing where the MiG had gone, the pilots turned their defenseless aircraft south and headed for the safety of Saudi airspace, leaving the F-15Es unprotected as they entered the dangerous skies of western Iraq.[141]

Unaware that his flight was not supported, Scott led his six jets toward their targets. Thirty miles from the targets they started seeing airbursts from radar-controlled AAA. At 10 miles out they came under attack from SA-2 and SA-3 radar-directed missiles. Corvette 01 and 02 made their attacks. As Corvette 03 approached the drop point, Eberly detected a warning from his radar-warning receiver that an SA-2 was tracking the aircraft. Almost immediately, he spotted a missile approaching from the right. He jinked into the missile and it streaked by. He had just rolled back to the left to release his bombs when the aircraft was rocked by a violent white explosion from a second missile.

The two men were momentarily stunned by the blast. Eberly tried to focus as he scanned the instrument panel but was quickly overwhelmed by the rapidly increasing number of warning lights. The aircraft had been fatally wounded. In the backseat, Tom Griffith tried to make a Mayday call on the radio. Instinctively, Eberly pulled the ejection handle, ejecting him immediately. Griffith ejected quickly thereafter.[142]

Both men floated down through the bitterly cold night air. Griffith landed uneventfully. Eberly lost his helmet and oxygen mask in the ejection. Aloft in the thin air, he succumbed to hypoxia and passed out. He woke up on the ground but was mentally confused. He had not taken any refresher courses in combat survival during Desert Shield. Additionally, he had not taken the time to develop a pre-mission evasion plan with the intelligence section back at the base. As his head cleared, he grabbed his parachute and moved away from his ejection seat, leaving the rest of his survival kit behind.

The two men could not see each other. Eberly took out his PRC-90 radio and made an emergency call, "This is Chevy . . . ." He stopped and remembered that Chevy had been his call sign on a previous mission. He started over, "This is Corvette 03 on guard. How do you read?" There was no answer. Sensing that his WSO was not too far away, he called Griffith, who also had a PRC-90. Visibility was good, so using various distinctive landmarks, they were able to rendezvous after only 15 minutes and immediately started looking for cover.

Griffith had all of his survival gear. Together they moved off to the southwest. As the sun came up, Griffith could see that Eberly had a gash in the back of his head and a bad scrape on his face. He tended to him as best he could, then they wrapped themselves in the parachute and got some sleep.

Corvette 01 reported to AWACS that Corvette 03 was missing. They immediately notified the JRCC. It had been a hectic night with numerous reports of aircraft down and emergency beacons detected. The JRCC controllers had to sort through the data coming in to them. Initially, they checked with all the airfields where an F-15E could have landed. When controllers could not account for the aircraft, they had to assume that they were down and then attempt to locate them. Returning to base, one member of the Corvette flight reported that the aircraft had gone down at approximate coordinates of latitude 34°13' N, longitude 040°55' E, about 10 miles southwest of the target. That was a start, but considering that it came from an aircraft moving at more than 500 miles per hour and under fire at the time, it was not accurate enough to launch highly vulnerable helicopters into such a high-threat area. Data came in from the SARSAT, but its CEPs were too large to commit any helicopters.

Colonel Hampton monitored the inflow of data and recalled, "We knew they punched out. We had intel on them from the RC-135 that the Iraqis were looking for them for a while and one ground group said that they had captured them."[143]

Hampton wanted to launch helicopters immediately, but he did not have the authority to do so.[144] Following the CSAR plan, they passed that data on to SOCCENT. There, Col George Gray began to intensively study the situation. He could stage MH-53s out of ArAr, and indeed, crews there were directed to begin mission planning. But they would need tanker support, which was problematic because that whole section of Iraq was one of the most highly defended areas in the entire country. Sending in tankers would put a large number of aircrew at risk.[145] The JRCC had not reported an accurate position on the men or even that they were alive and still free. Gray would not send his helicopters in to do a search. It did appear that the men, if they were there, were above the dividing line with the combat rescue forces in Proven Force. He suggested to his boss, Colonel Johnson, that they pass the mission to them.[146] They sent an alert message to the guys in Turkey, who began initial planning.[147]

Throughout the night and next morning, the JRCC worked with various intelligence assets to determine the status and position of the survivors and alerted them to be vigilant for any radio calls from the two men. But such signals were dangerous because the Iraqis were known to have excellent homing capabilities to track any calls from the PRC-90 radios. Intelligence sources also suspected that the Iraqis themselves were setting off beacons as decoys.[148] All of this had to be resolved.

Additionally, General Glosson indicated that some things were going on behind the scenes. Other government agencies were working to get the two men, possibly by passing "operatives" through Syria. Such a plan was discussed at SOCCENT, and the JRCC guys were aware that some possible options were being considered.[149]

Colonel Gray suggested to Colonel Johnson that they request access through Syria for a possible helicopter mission. They sent it up the line, and CENTCOM made a formal request through diplomatic channels for authorization to use Syrian airspace. They expected approval since Syria was an ally in the war.[150]

As Eberly and Griffith slept, intelligence sources picked up what appeared to be another downed aircraft in the area. They monitored a Mayday call for someone using the call sign Crest 45A. The JRCC quickly did a check with the air operations center and determined that there was nobody using that call sign. They also checked with the Proven Force command center and got the same result.[151]

Meanwhile, a civilian had come into the US Embassy in Amman, Jordan, and said that he had information that Eberly was alive. He suggested that he could turn him over to the American government for a reward. This took time to resolve as bogus.[152]

As time passed with little apparently happening, some personnel in the TACC began to question the lack of activity on the part of the rescue forces. Capt Randy O'Boyle began to take some heat from some of the fighter guys. He recalled, "Some of the F-15 guys were giving me some [heat]. I said, 'Look, next time you're out there . . ., why don't you just descend down to altitude, drop your gear, drop your flaps, and just go look for somebody on the ground in one of those spots where you think there is somebody. And as soon as you get him then you let us know. . . . You just can't go trundling into some place in a high-threat environment without knowing exactly where the guy is."[153]

His view was shared by Col Ben Orrell, Colonel Gray's director of operations, back at the 1st SOW at Hurlburt and SOCCENT. Colonel Orrell said,

> I'm not kidding you, you could see those Paves [MH-53 Pave Lows] 50 miles off. There was no hiding them. That's a big ol' slow moving target—I was reluctant to go cruising in there in the daytime. There certainly may have been a situation where we would have done it, but if you don't have a guy talking to you on the radio, it's pretty hard to convince me to send another two or three crews in there. . . . The only way we were going to survive as a rescue force in that environment was to fly at night. And I don't think that the fast movers [fighter pilots] ever accepted the fact that we were not just going to come plunging in there in daytime like we had done in Vietnam. Had we done that, we'd have lost more [crews].[154]

But that was exactly what the F-15E crews expected. When Eberly did not return, his assistant, Col Bob Ruth, became the acting director of operations for the wing. He had flown OV-10s over the Ho Chi Minh Trail and recalled, "When we were over in SEA [Southeast Asia], if an airplane went down, we dedicated just about any air we could to try to suppress the area to try to get the survivors out."[155]

But that was not happening here. The F-15Es were held for the anti-Scud missions. Colonel Ruth said, "Everybody just kept doing his mission and everything was handed over to the [C]SAR folks."[156]

During the day, several strike flights flew through the area. As each would pass, Eberly or Griffith would attempt to make contact. One of the groups was a flight of four F-16s from the 10th TFS led by its commander, Lt Col Ed Houle.

As with Corvette the previous night, his mission had been a last-minute change from a 40-aircraft package of F-16s, F-15s, and other assorted support aircraft against a target in the Baghdad area to the Scud sites out west. In preparation for the new mission, he had not been briefed that any crews had been shot down in the Al Qāim area. Approaching his target he heard, "This is Corvette 03. Does anybody read?"

Not expecting such a call, he let it pass and struck his target. That fully taxed him since it was well defended and the AAA and SAM sites were active. Leaving the area, he checked out with the AWACS. He asked them, "Hey, who is Corvette 03?"

The controller responded, "Why, did you hear something?"

Houle replied that he had. The controller thanked him and directed him to return to his base.

Landing back at his base and now fully suspicious, he walked into intelligence for the mission debrief. "Who . . . is Corvette 03?" he demanded. The intelligence officer informed Houle that Corvette 03 was an F-15E shot down in the Al Qāim area the previous night and that rescue forces were trying to locate them for rescue. Houle's cockpit mission tape was pulled and forwarded to the 4th Wing, where unit members identified the voice as that of Colonel Eberly.

All of this came as a complete surprise to Houle because no one in the flight had been briefed that there was a downed aircrew in that area. If they had, they would have been on the lookout for them. Houle made sure pre-mission briefing procedures were changed to ensure that all of his pilots were briefed on downed crews or CSAR efforts going on in any areas in which they would be flying. In fact this was passed to the JRCC, and they then sent out a report every 12 hours to all units giving information on downed aircrews.

Houle also recommended to higher authorities that all aircrews should use the term *Mayday* if they were down instead of just talking on the radio. That would get the immediate attention of any aircrews listening.[157]

That evening, as another flight of F-15Es was passing through the area, one of the crews, Chevy 06, apparently made momentary voice contact with Griffith. Chevy 06's position at the time was almost 30 miles southwest of the target area and did not correlate with any other reports. The flight lead on this flight was Griffith's squadron commander, Lt Col Steve Turner. He was certain that he had talked to Griffith.[158]

Returning from his mission, Turner called and spoke with some of the controllers in the JRCC. They asked if he had authenticated either man by asking any of the personal questions each kept in his file for such occasions. Turner replied that he had not, but he was absolutely certain that it was Griffith on the radio because he had served with the man for three years. He became adamant that more be done to get the men out. He was coming to the opinion that the rescue forces were slow-rolling them for some reason. When the JRCC explained the difficulties they were having locating the guys, he suggested, strongly, that an F-15E be sent in, not as part of a strike package but specifically to find the guys.[159] He was told the F-15Es were needed to hit targets. CENTAF would send in F-15Cs to search.[160]

The 4th Wing commander was also upset. He called the JRCC and is quoted by one of his young majors as having said, "Is this incompetence or is it just sheer cowardice?"[161]

Tension was building in the TACC to do something to rescue the aircrew. Some wanted to send the helicopters regardless; others were more cautious.

Colonel Hampton took direct action: "Every mission that went up into that area we tasked to make radio calls, to monitor the frequencies. We tried everything that we could in order to make contact with those guys."[162]

Word of the lack of action spread through the 4th Wing. Another F-15E pilot, Maj Richard Crandall, said, "You can't believe how angry we were that they were not going up there looking for those guys. We were so angry that Al Gale and I actually proposed to take a vehicle and drive up there to get them."[163]

Col George Gray was holding the line. He was not sure where the two men were. He also knew that without good authentication that the voices on the radio were, in fact, Eberly and Griffith, it could be a trap. The North Vietnamese had done this numerous times. He did not want to lose a helicopter crew or fighter escort to a trap. Solid procedures had been established to authenticate downed crewmen, but they were not being followed. He had no control over that. That was JRCC's business. He did not want to commit a helicopter crew until he was sure. But the pressure was building to do something.[164]

By this time, Colonel Hampton was not sanguine about the chances of recovery or even the efficacy of making an effort. He recalled,

> The guys could have still been down there, and if you can do it without losing anybody to do it, great. But as far as pushing for a mission at that point, we weren't in that position, and we left it up to the SOC[CENT] guys to determine. If you want to go in there and do it, fine, okay—but we didn't agree on coordinates. We had a position that was farther to the east. Why they went to where they did I think was based on some cuts from an RC-135. I'm not sure that was a real good position on the guys. That was probably two-day-old data at that time, and if your RC-135 is down here [in Saudi] and you're doing DF up to that position . . . I know their gear is sophisticated, however there's got to be some precision there.[165]

Up at Batman in Turkey, the Proven Force MH-53 crews were busy planning contingencies. They were collecting all available information, although this was a difficult task at such a remote location. Given the general area of the survivors, they could see immediately that flying east into Iraq and then south to the Al Qāim area was much too dangerous. One of the pilots, Capt Steve Otto, explained the tactical problem. "[Flying east] we simply would not have the range to make it down towards Al Qāim. So given the threat and that large obstacle in the tri-border area . . . we knew that if we were going to get to Al Qāim, we were going to have to fly over Syria. We started asking for over-flight permission to go through Syria."[166]

That was not all that concerned the Pave Low pilots. Captain Otto explained,

> We had a call sign but we did not have any survivor data or any ISO-PREP [isolated personnel report] information. Perhaps more disturbing was that we had three last-known positions and they were in a triangle that was about 20 miles on each leg. The bad part about it is that, had it been in a low-threat area, it would have been no big deal and we probably would have been less intimidated. But this area, Al Qāim, had intense AAA and surface-to-air missile defenses. We knew that that was one of the early target areas and Corvette 03 had been shot down striking targets in that vicinity . . . this was really an intensely defended area.[167]

Back at Al Kharj, Col Bob Ruth listened to the tape supplied by Ed Houle and then had it flown to the JRCC. There, Hampton and the JRCC controllers listened to it and the pleadings of the guys from the 4th Wing. They called Colonel Gray at SOCCENT. Gray was finally convinced by their assurances and acquiesced. With positive voice contact, SOCCENT decided to launch a mission.

They formally tasked the combat rescue forces in Turkey. There was enough darkness left that they could make it in the dark—their preferred method of operations. But there were other issues with which to deal. First, the pilots, Capt Grant Harden and his copilot Capt Matt Shozda, and Capt Steve Otto and his copilot Capt J. D. Clem, all felt that they did not have enough solid intelligence data to properly plan and execute the mission. They took the matter to their squadron commander, who went to work to get them better information, especially a more precise location for the survivors.[168] Second, Turkey was still skittish about the entire Proven Force operation and refused them launch authority. By the time all of that was resolved, they had lost the night.[169]

Getting into the Al Qāim area in a helicopter was a tough tactical challenge due to the level of enemy defenses. George Gray had already concluded that any approach from the south with his helicopters and their tankers was just about suicidal. The problem was similar from the north. Any approach down out of the mountains along the Turkish-Iraqi border and then across the flat midland of Iraq was just as dangerous. But an approach through Syria looked much safer, as Colonel Gray had recommended at the beginning.

Syria had not granted a flight clearance for the mission. Instead, the Syrian authorities recommended that they themselves send in a team to pick up the two flyers. Then to further confuse the entire issue, that Bedouin tribesman had come to the US Embassy in Jordan claiming to have a "blood chit" from one of the flyers. He wanted to trade the two men for a new truck.[170] All of the political wrangling just cost further delays.

The next day, Eberly's spirits soared when he heard what was obviously a combat rescue effort under way. He began calling on the radio, but could get no response. In fact a rescue operation *was* going on; it just was not for him and his WSO, but for a Navy F-14 crew (Slate 46) who had gone down well to the east.

At the JRCC, Colonel Hampton was trying to task every aircraft going into the Al Qāim area to listen for and try to locate the two men.[171] Unaware of the efforts being made in their behalf, Eberly decided that they were not going to be rescued. He talked it over with Griffith, and they decided that since they were so close to Syria, perhaps they could walk out. They had a map, although it was out of date, and felt that they had a fairly good idea of where they were, so they set out walking west. They encountered some Bedouin camps and even some vehicles but were not detected.

That evening the pilot of an F-15C, call sign Mobil 41, heard their emergency beacon and directed them to go to the backup frequency.

Switching over, one of the men said, "Go ahead."

Mobil 41 responded, "We are just trying to get hold of you to see if you're still around. What is your physical condition?"

Griffith responded, "Physical condition is good. Alpha and Bravo are together. We are approximately 10 miles northwest of [garbled]."

Mobile 41 responded, "Corvette 03, we read you. We will be flying closer to get better radio contact."

Anxious, Eberly asked, "Do you understand our position?"

At that point, somebody came on the frequency and shouted, "SAR in effect, get off of this frequency." The interloper did not identify himself. But Mobil 41 could not reestablish contact with the crew of Corvette 03 and was not able to verify their position or authenticate them. The two men continued to call on both frequencies, but their calls went unanswered.

Moving to the west, the two men discussed pressing on to Syria or hiding and waiting for rescue forces. Near the border, Eberly spotted what appeared to be an abandoned building. Both men were cold and soaked, and the idea of being inside, sheltered from the wind was appealing. Eberly tiptoed up to the building and looked through one of the windows to see if it were safe. The building suddenly exploded with automatic weapons fire as a dozen soldiers came running out. In the early morning hours and only a few miles from the Syrian border, the two Americans were captured.[172]

The rescue forces did not know of the capture, but did notice that they no longer had contact with the two men. Throughout the day, strike flights hit targets in the Sam's Town area and called for Corvette 03. Intelligence assets kept a tuned ear for any sign of the men. Nothing was heard. The SOCCENT commander conferred with the SOCEUR commander about the mission. They reviewed all pertinent known data on the two men. Finally satisfied that the mission had a reasonable chance of success, Col Jesse Johnson, the SOCCENT commander, directed them to execute, contingent upon Syrian approval to use their airspace.

The men from Proven Force were primed and ready to go in that night. Again, Capt Grant Harden and Capt Steve Otto and their designated copilots would fly the two MH-53s. This time, supplied with better intelligence and necessary data on the survivors, they planned to make the flight through Syria escorted by an MC-130 tanker aircraft. Their arrival in the Al Qāim area would be coordinated with several air strikes designed to divert the attention of the SAM and AAA sites. SOCCENT and the forces at Batman felt that the supporting air strikes and the use of Syrian airspace would give the best chance of success and survival of the rescue helicopters and crews. But copilot Clem was dismayed that even as they were being driven to their aircraft, they were still being given "updated" coordinates on the survivors. And they were varying by as much as 20 miles.

At launch time the Syrian approval had still not come through. Captains Otto and Harden with their crews were ready in the massive MH-53s. On board each helicopter, in addition to the pilots, were two flight engineers, two door gunners, two PJs, a combat controller, and a few Army Special Forces troops for ground security. They reported that they were ready and were directed to launch. They started engines, and Captain Otto discovered that the radar jamming system on his aircraft was inoperative. He decided to go anyway, and the flight took off.

Approaching Syrian airspace they still had not received the necessary overflight clearance. The command center told them to press on to the objective area. Twenty minutes later they were notified that approval had been obtained from Syria to enter its airspace. En route they received several more calls on their SATCOM radio reconfirming their clearance. The numerous calls became a distraction.

For two hours, they flew toward Al Qāim. Their flight path was right along the Iraqi-Syrian border. They tried to maintain an altitude of about 100 feet. There was no moon, but starlight illumination was enough for night vision. The pilots found the flying to be challenging. They had not gotten used to flying blacked out over shifting sand dunes and could not help but become concerned every time the radar altimeter showed that they were less than 10 feet off the ground. Several times they had to climb abruptly to avoid towers and wires, and at one point, they flew through a flock of birds. Several hit each aircraft.[173]

Approaching the Euphrates River, they were locked on and tracked by an SA-6 from their right. It was obviously a Syrian site, and they moved to avoid it. With an inoperative radar jamming system on their aircraft, this especially concerned Captains Otto and Clem, but they gave no thought to turning back.[174]

Capt Grant Harden had a solid plan. "Our plan was to hit a final IP [initial point] and then make a run-in. The run-in to the exact location would be based on contact. If there were no contact, we would not go beyond the final area."[175]

As the mission proceeded, it was being monitored nervously, even by commanders back in the United States. Brig Gen Dale Stovall at AFSOC said, "We held our breath. There was a tremendous amount of pressure to send [the helicopters] in to search when we didn't have a good fixed position on those guys."[176] He was about to go crazy because it brought back powerful memories of Jolly Green crews sent in to North Vietnam to look for downed fighter crews in 1972. The losses suffered by the rescue crews back then had been high. Stovall could still see their faces.

Near dangerous Al Qāim, the two blacked-out helicopters entered Iraq. Low and slow, they began to move to the IP. Captain Harden had requested that they be supported by preceding air strikes against Iraqi SAMs and AAA in the area and one fighter aircraft to act as on-scene commander (OSC). Specifically, the designated OSC would arrive just ahead of the helicopters, contact the survivors, authenticate them, and have them ready for a quick pickup. But it didn't turn out that way. Harden recalled, "The entire sequence, as always happens, did not come off as planned. There was supposed to be a diversionary covering air strike. It was late and short. When we went in, we were supposed to have a high bird make contact. That never happened."[177]

Without an OSC, they were basically on their own. Captain Otto described how the mission proceeded:

> We got down to the [IP] and we started holding in kind of a figure 8, not to fly over the same ground track. We were about a mile into Syria. Our ROE [rules of engagement] from our squadron commander was that we were not going to fly or commit into the threat area around Al Qāim unless the survivors came up on the radio. We noticed that there was a slight rise and we could stay masked. As all of this is going on, the giant light show is going on with the AAA. It was towards the strike aircraft but randomly. It wasn't guided toward us. It was just fired up into the air, which is kind of the way they seemed to do things. We eventually got there and realized that the fighters were not going to get Corvette 03 up on the radio.
>
> We orbited for about 5 minutes and expected to hear them call. We were on time as we got to the orbit point. It coincided perfectly with the strikers getting there.
>
> Then eventually, Grant and his copilot Matt Shozda realized that it was just getting screwed up and we were going to have to do our own authentication. And Grant told us to stay down low. He climbed up a couple of hundred feet, maybe 500 feet, and just started talking on the radio . . . trying to get them up on the radio. Probably after about a minute delay, we started to notice that the Iraqi AAA started to get real intense, once we had talked on the radio. And even in the aircraft, we felt that they were intercepting and DF-ing us. Then the fighters joined in trying to get them up on the radios. Corvette 03 only had PRC-90s. And we knew as long as we were there and talking on the radio, the

odds of the mission being compromised were greater. Bottom line is that we stayed down there for almost 30 minutes orbiting and calling on the radio. Never heard a word from Corvette 03. Then, reluctantly, without radio contact, we were done. We flew back to Turkey.[178]

Capt Matt Shozda in the other aircraft had similar memories:

> I got on the radio and started trying all the different frequencies to contact him. Somewhere at that point, we realized that the SAR net was nothing more than a radio-controlled AAA, pilot-controlled AAA. We would key the mike and they would start firing. I told Harden, "Look! They're DF-in' us. Watch this!" So I made a radio call and they started shooting again. He told me, "Cut that out!" They definitely had a trap set up for us. They were waiting for us because the final location that we got—that's where all the AAA was coming from.[179]

With no contact, the two crews left the area according to plan. Arriving back at Batman, they went into crew rest. The next day they were back into the alert cycle. Within a few days their unit at Batman established a communications link with higher headquarters, which enabled them to get daily intelligence updates and even the ATO [air tasking order]. Whenever coalition crews were flying over Iraq, MH-53 crews at Batman were on alert for combat recovery. It was their primary mission.

The next day General Jamerson visited the base and was briefed by the crews on the mission. He listened intently and then declared that searching for survivors in high-threat areas was a task for fighter crews, *not* helicopter crews. The MH-53 crews concurred.

A few days later the MH-53 crews at Batman got a chance to see the CNN footage from Baghdad of the first night of the war. They could not help but notice that the AAA was much less intense that what they had seen at Al Qāim.[180]

For the next several days, flights into the Al Qāim area continued to listen and call for Corvette 03. Not ready to give up, Colonel Gray had his forces build a task force of MH-60s, MH-53s, and MC-130 tankers at ArAr for another attempt. The aircrews planned diligently for the mission and waited for the execute order, but they were under no illusions as to the difficulty of reentering the Al Qāim area.[181] But Gray never sent the execute order, because neither the location nor status of the two men could be determined. The reason was simple—Eberly and Griffith were on their way to Baghdad.

But the saga of Corvette 03 was not over. The men of the 4th Wing were bitter about the nonrescue of their mates. Their bitterness peaked a few weeks later when they saw Eberly's face on CNN—as a POW. It hurt to see him. To a man, they had all been more that ready to help out in the rescue effort. Col Bob Ruth said, "If they [JRCC] had called down and said 'Hey I need a 4-ship, can you round up enough people?' we would have had people cooking. We had the planes. We had the people. We could easily have done along those lines without impacting the ATO. But we were never asked."[182]

One of the unit pilots expressed the feelings of most when he later wrote, "Our DO [director of operations] and backseater were on the ground for three and a half days in western Iraq. Nobody would go in and pick them up, and they eventually became prisoners of war. Before the war, the Special Operations guys came down to talk to us. 'No sweat,' they said, 'we'll come and get you anywhere you are.' That from my perspective was a big lie. . . . Nobody was going to come and get you."[183]

In an after-action report, the 4th Wing commander himself said:

> It seemed to me that the forces running the SAR wanted a perfect situation. Before they would launch they wanted to know exactly where they [Eberly and Griffith] were, that they had been authenticated and on and on. I mean, when we got the tape I had [Lt Col] Slammer Decuir, Griff's roommate . . . listen to the tape and verify that it was Griff. But those guys at JRCC would not take our word for it. So we fly the tape to Riyadh and they say, finally, "Yep now that we have heard the voice, we believe what you heard is in fact true." I mean it was frustrating beyond belief that we had to prove to others that, yes, there were people out there who needed to be picked up. What frustrated me the most was that I couldn't push the right buttons to get the SAR going. Horner and Glosson, my bosses, would have broken their necks to get up there, but they were running the air campaign and had no control over the SAR effort.[184]

General Glosson was also frustrated by these events. He and Captain O'Boyle had some heated discussions about the SOCCENT response. O'Boyle repeated that there were places that helicopters could not safely go. Glosson said,

> Randy [O'Boyle] is 100 percent correct on that issue unless I made the decision I was willing to lose them. If I'm willing to lose them as the commander, I should have the prerogative to send a helicopter or send two or three, understanding I may lose one of them. That's my decision. It should not be someone else's decision. I am not saying you send people

into harm's way just to say you did it. But many times . . . you can assist the CSAR effort with distractions in a way that a helicopter can sneak in and not have near the exposure. During Desert Storm, AFSOC [SOCCENT] wanted to look at everything in isolation. They wanted to say, "Oh, a helicopter can't get in." Randy and I had a few conversations on this. I said, "Randy, stop letting those guys look at this in isolation. I can make all hell break loose a quarter of a mile from where we want to pick the pilot up. I can make sure the people on the ground are only concerned about survival." Bottom line, you can't look at CSAR or anything else during a war in isolation.[185]

Neither Horner nor Glosson could launch the rescue helicopters across the FLOT [forward line of own troops] because General Schwarzkopf had given that responsibility and authority to the commander of SOCCENT, Col Jesse Johnson. Johnson's air commanders, well schooled in the realities of rescue behind enemy lines, delayed the effort until they felt that their crews had the best chance of rescuing the men and not being shot down themselves. It was an unfortunate misunderstanding fueled by the fog of war. The fighter guys expected to be picked up. For years, they had heard the motto of the rescue forces, "That Others May Live." They *knew* that the rescue guys would come. But when the helicopters did not, for reasons that they could not know or even understand, they lost faith and condemned those responsible.

Yet, Col George Gray was adamant in his logic. "I wasn't going to send guys into a situation where we were automatically going to lose a helicopter and 5 more guys."[186]

At CENTCOM, Colonel Harvell watched all of this and was dismayed at the attitude of the Air Force officers in their enthusiasm to so quickly send the SOF helicopters into such a high-threat area. He said, "This is an issue of recurring special operations concern in that non-SOF people have a tendency to commit special operations forces in unrealistic ways. This is a theme that the special operations forces guys have got to fight. They [Air Force] say 'Send the special operations forces guys.' It's not a special operations forces mission. It's easy for them to say, 'Mount up the special operations forces guys and have them do this.' They don't understand what our strengths and weaknesses are and what we can and cannot do."[187]

Concerned about the failure, the commanders in CENTAF tried to address the problem. Analysis indicated that the PRC-90 radio used by the crews was clearly inadequate for the conflict. It had only two frequencies available and was too easily exploited and compromised by the Iraqis. The newer radio, the PRC-112, was only available in limited numbers. It had the ability to load more frequencies and had some discrete homing capability. Prior to the war, the special operations forces and the Navy had bought more than a thousand for their troops; the Air Force had not. Realizing their mistake, the director of operations for CENTAF sent a message to the Pentagon asking for some help. Specifically, it asked for:

1. Several hundred of the PRC-112 radios for aircrew use.
2. Modification kits for more homing receivers for helicopters. (The MH-53s and Navy HH-60s were already modified.)[188]

Interestingly, the message did not ask for modification kits for any fighter aircraft, not the F-15s or the Sandy A-10s.

Perhaps a better choice would have been to modify the 72 Block 40 F-16C/Ds, which had GPS. Modified with the DALS, they could locate survivors by homing on their PRC-112s, then pass the coordinates from their GPS to the MH-53s for precision guidance.[189]

Although never stated explicitly, it seems that these aircraft were needed for other missions—hunting Scuds and destroying Republican Guard divisions. As incredible as it seems, the scheduling of aircraft was extremely tight. As one scheduler noted at about this time, "With all the aircraft available in-theater, I found it difficult to believe that we were actually 'short' [of available aircraft to strike the Scud sites]. We did, however, have that problem. With the number of packages and individual missions scheduled in the ATO, there are, in fact, few unscheduled aircraft available."[190]

Colonel Ruth and his men were ready to fly, but unlike the days of the Bengal 505 SAR in Southeast Asia when almost unlimited sorties were available, there were limits on the number that could be produced in Desert Storm. Almost every sortie had a designated target as part of the campaign plan.

None of that postmortem helped the crew of Corvette 03. Colonel Eberly and Major Griffith were POWs, and the war went on.

The strategic immediacy of the hunt for the Scuds continued to call for the diversion of air assets. There were also Scud launch boxes in the east, and the Block 40 F-16s, along with the F-15Es, were diverted to that mission. At times, as many as four aircraft would maintain an airborne presence over the sites with up to eight more on ground alert. A-10s were with them prosecuting the same mission.[191]

General Glosson had another asset available to him that had the ability to loiter over an area for long periods of time and bring great firepower to bear. That was the AC-130 gunships of the 1st SOW. They had deployed with the other AFSOC forces in August and September but, unlike other AFSOC assets, they were being directly fragged by CENTAF. This was a real point of contention between Col George Gray and the CENTAF commanders. Simply put, Colonel Gray was concerned that the gunships were being given missions for which they were not suited—specifically, that they were being sent into high-threat areas where they were too vulnerable.

The AC-130 was a powerful beast. Introduced during the war in Southeast Asia, it ruled the skies over the Ho Chi Minh Trail. The North Vietnamese called it the "Thug" and feared only the B-52s more. With its array of sensors and 20 mm, 40 mm, and 105 mm guns, it could inflict serious damage upon any force or target. But *any* C-130 is a big, fat target to enemy guns and SAMs. Gray did not like the way they were being used.[192] This disagreement added to the disgruntlement building between the CENTAF officer and the SOCCENT guys.

20 January

During the early morning hours, a small force of British commandos from the Special Air Service (SAS) crossed the Saudi-Iraqi border into western Iraq. This operation had been recommended to General Schwarzkopf by the commander of the British forces in the Gulf, Lt Gen Sir Peter de la Billière. Their initial mission was to harass Iraqi forces in the area, distract the attention of their leaders, and siphon off forces from elsewhere in Iraq. On the ground, they began attacking military units, convoys, and truck traffic along the main east-west highway. But by the third day, they had been retasked to searching for and attacking the Scud sites and missiles. Eventually, this force would consist of several eight-man teams and four 30-man teams driving heavily armed four-wheel vehicles and motorcycles.[193] Ultimately, 250 SAS troops would be involved.[194] One of the teams was inserted near the wreckage of the F-18 flown by Commander Speicher, lost on the first night.[195]

Procedures were developed for the SAS teams to contact and direct the F-15Es and A-10s being diverted into the area to attack the Scud sites. It is conceivable that they could also have been used to recover the crew of Corvette 03, but no records exist in any unclassified sources that such a mission was ever contemplated.

The effort to find and kill the Scuds was growing. By 24 January, CENTAF, at the direction of CENTCOM, had diverted 40 percent of its air strikes to the hunt for the Scuds.[196]

Seeing the insertion of SAS teams into the war, the commander of SOCOM, Gen Carl Stiner, began to lobby for a mission there for some of his troops. General Schwarzkopf still had an abiding distrust of commando operations, but after the Israeli government suggested that Israel be allowed to insert commando teams into western Iraq, Schwarzkopf relented to Stiner's request and agreed to a SOCOM task force under his direct control.[197] Meanwhile, the air campaign continued unabated, and coalition aircraft began to suffer from the wear and tear of intense and sustained operations.

I030/M005 – Stamford 11/Tornado, RAF

A Tornado flown by Sqn Ldr Peter Batson and Wg Cdr Mike Heath had flight control problems upon takeoff.[198] The crew jettisoned their bombs and orbited for more than an hour to burn off fuel for an emergency landing. Continued flight control problems made a landing impossible, and the crew ejected. With the assistance of a crewman from HCS-4 using NVGs, a Royal Saudi helicopter recovered the crew.[199]

This was not a CSAR, but the JRCC got involved in cases like this because they had responsibility for rescue throughout the AOR. Fortunately, the Saudis had an excellent fleet of helicopters that was scattered in detachments throughout the kingdom. Whenever something happened that required a local response, they were the asset of choice. Even in times of war, these missions occur and must be flown.

In directing the air campaign, the CENTAF commanders had another problem. The Iraqi Air Force had not reacted as expected. American planners thought the Iraqis would launch their fighters to oppose the air assault and be blown away by the hordes of F-15s and F-14s dispatched to sweep the skies ahead of the attack aircraft, but this had not happened for several reasons. First, bad weather over much of Iraq had slowed the air campaign. Second, the Iraqis had sent some fighters up against the coalition, and numerous kills had been logged by both Air Force and Navy fighters. The Iraqis quickly learned the lethality of US forces.

After three days, enemy activity had dropped off considerably. The Iraqis seemed to be sheltering their aircraft for some future use. Still, they needed to be destroyed lest they be used at some point in a way that could catch the coalition forces off guard. The answer was to go after them on their airfields and in their hardened shelters.[200]

I031/M006 – Slate 46/F-14, USN

The crew of LT Devon Jones and LT Larry Slade launched on their first combat mission to escort an EA-6B supporting an early morning strike package hitting the Al Asad Airfield just west of Baghdad.[201] Entering the target area with clouds layered to above 20,000 feet, the aircraft set up their orbits, and the EA-6B launched a HARM against a preplanned target. While holding at the orbit point, Slate 46 was targeted by an SA-2 missile battery. Jones took evasive action, but the missile detonated just behind the tail, fatally wounding the aircraft. The F-14 entered a flat spin. Unable to control it, the crew ejected at about 0320Z.

As the two aviators floated to the ground, the crew of the EA-6B notified AWACS and gave their approximate position. Mistakenly, the EA-6B crew reported that the downed aircraft was an A-6. Then the JRCC was told that both an A-6 and an F-14 had gone down. It took some time for the JRCC to determine that the A-6 in question had diverted to an airfield in Saudi Arabia, but that the F-14 had, in fact, been downed.

Descending, Lieutenant Slade took out his PRC-112 survival radio and made an emergency call. One of the controllers on AWACS acknowledged it. Lieutenant Jones made an emergency call on his PRC-90 radio as soon as he reached the ground.[202]

Upon landing Slade moved quickly away from the wreckage. He attempted to make several more radio calls. None was answered.[203] He had observed his pilot once during the parachute descent but was unable to hear him on the radio either. The area was almost perfectly flat and bare. With no place to hide, he tried to dig a small trench, but the ground was just too hard. He tried repeatedly to make radio contact with anyone but was unsuccessful.

About 45 minutes later intelligence reported that Iraqi helicopters were apparently in the area looking for the survivors. The JRCC coordinated with the TACC to get a flight of F-15s into the area to provide top cover for the survivors. Within minutes they were overhead but made no contact with the enemy helicopters.[204]

The JRCC contacted SOCCENT and tasked them with the recovery. Initially, they repeated the erroneous report that both an A-6 and F-14 were down. SOCCENT passed the missions to the command center at ArAr, where rescue crews on alert began mission planning. The SOC commander there, Colonel Garlington (also commander of the 55th SOS), had MH-60s, Navy HH-60s, and MH-53s available. He put them on alert in that order, since the MH-53s had led the mission for Stroke 65. Since it was Navy aircraft reported down, the crews from HCS-4/5 wanted the missions and began intensively preparing for them.[205]

The weather at ArAr was terrible due to fog; visibility was measured in feet. Neither the MH-60s nor the HH-60s could safely take off in such low visibility. The MH-53s with better flight instruments could do so, and the missions were assigned to them. Capt Mike Kingsley and his crew would fly the A-6 recovery, and Capt Tom Trask and his crew would do the F-14 crew pickup.[206]

Both crews began intensive preparations. Subsequently, Kingsley was notified that the report of a downed A-6 was incorrect. He and his crew went back on alert.

Photo courtesy of Peter Mersky

The crew of Slate 46, Navy lieutenants Larry Slade and Devon Jones, had their F-14 shot down while escorting a strike against the Al Asad Airfield west of Baghdad.

With voice contact established with the pilot Jones, a reported location, and intelligence showing that the area around the survivors was not high-threat, Moccasin 05 commanded by Capt Tom Trask took off at 0505Z to recover the two-man crew of Slate 46.

As they launched, the command center at ArAr notified the JRCC that their estimated time of arrival at the survivors' location was 0630Z.[207] The helicopter lifted off and was immediately swallowed by the thick fog. Heading north, they flew directly over an Iraqi outpost just across the border. Within a few minutes, Trask and his crew began to get warnings that Iraqi SAM and AAA sites were going active in their area.

Trask knew that Slade had a PRC-112 radio. On board the helicopter, his crew had the necessary codes to precisely home in on his signal if they could get close enough to pick it up, but it would be hard to do at the extreme low altitudes at which they needed to fly for protection from the Iraqi guns and SAMs.

The JRCC also had A-10s on alert for CSAR at King Fahd and KKMC. That morning two pilots, Capt Paul Johnson and 1st Lt Randy Goff, were slated to take off from King Fahd to fly to KKMC and assume alert as Sandy 57 and 58. They were delayed for a short while waiting for the morning fog to dissipate. Getting airborne they were instructed to contact AWACS for divert to Slate 46. After working their way through thick clouds and around thunderstorms, they were able to rendezvous with a KC-135 tanker to top off their fuel and then check in with the AWACS.[208]

Based upon the estimate from Moccasin 05, the JRCC coordinated with AWACS to divert more F-15s and two flights of A-10s—call signs Springfield 17 and 18—and Sandy 57 and 58 into the area to fly CAP for the helicopter. Below, Trask followed the best spider route into the area. AWACS gave them threat warnings as necessary as they moved ever deeper into Iraq. Thirty minutes into the flight, the clouds started to thin a little. Visibility was improving, and they were able to see the ground. The terrain was barren and almost completely flat.

Inbound to the reported positions of the two men, Trask received a warning from AWACS that Iraqi MiGs had taken off and were heading toward his aircraft. Trask was holding just below a low deck of clouds. He descended and hovered his helicopter inside a wadi to take advantage of whatever concealment it offered. AWACS called that the bandits were passing directly overhead and immediately vectored F-15s against the intruders, who prudently fled. Neither Trask nor his crew ever saw them, but when the threat was clear, he resumed his low-altitude route.[209]

Then AWACS reported that four Iraqi helicopters had taken off from a nearby airfield and were heading toward them. At one point, the enemy helicopters were broadcasting on the emergency frequencies saying, "This is rescue, American pilot; this is rescue, come up voice."[210]

The enemy helicopters did not concern Trask since he was sure he could defeat them in an air-to-air engagement. Before he could confront them, they turned around and landed back at their airfield.

Approaching the reported area of the survivors, the crew of Moccasin 05 began searching for the two men below and the F-15s above. They could not establish contact with either. An EA-6B that was part of the original strike package was still orbiting in the area and reported that the survivors were actually 50 miles north and that the initial coordinates reported by AWACS were terribly inaccurate. This caused some discussion between Trask and his crew and the headquarters at ArAr as they pressed SOCCENT for location data on the survivors.

The terrain in the area was almost completely flat. To avoid the enemy defenses as much as possible, Captain Trask kept his aircraft low and pressed on. The weather had cleared out considerably, and the visibility was now about 30 miles. The crew could not see any enemy activity or even any settlements nearby and felt relatively safe.

Flight at low altitude, though, reduced the range for the helicopter crews to hear the survivors' emergency radios. After searching unsuccessfully for 25 minutes, Moccasin 05 turned for home base. As it was leaving, one of the A-10 pilots in Springfield flight contacted Trask and reported that he thought he had made radio contact with a survivor. Trask returned to the area, but this effort was also unsuccessful, and Moccasin 05 had to return to ArAr for fuel.[211]

At approximately 0730Z, Lieutenant Slade spotted a white Datsun truck. Inside were two civilians, both armed with rifles. They saw Slade and took him prisoner.[212] He was not able to communicate to the rescue forces that he had been captured. Consequently, the rescuers remained on the lookout for anything that might signal the status of Lieutenant Slade.

Lieutenant Jones never heard his mate on the radio. He also moved away from the burning wreckage, initially east, and then northwest. He observed many tire and foot tracks. Walking for quite a while, he finally found a small mound that afforded some protection. Not too far away was a small wadi with a blue water tank. Using his knife, he dug a small trench in the hard ground and tried to cover himself as best he could.

Remembering what the rescue briefers had said, he did not expect a rescue attempt until night. At about the same time that Slade was being taken captive, Jones watched a small truck pull up to the tank. He sprinkled sand on himself to try to conceal his dark flight suit and boots. A few minutes later the truck drove off. Relaxing a little, Jones noticed a black scorpion crawling on his sleeve. He jumped out of his hiding place, flung the scorpion to the ground, and killed it.

High above, the AWACS controller working the mission instructed Sandy 57 and 58 to proceed directly to the survivors' estimated position. As they approached, the controller told them that the rescue helicopter had been unable to locate the survivors and had returned to base for fuel. He also gave them new instructions to head southeast and search for some suspected Scud missiles. Assuming that the CSAR had been terminated, they reluctantly complied. After 90 minutes of fruitless Scud searching, they rendezvoused with a KC-10 and took on another load of fuel. They checked back in with the AWACs, fully expecting to be sent home, but were diverted back to the CSAR effort.

Captain Johnson acknowledged his instructions and once again turned his flight north. As they flew into the estimated area, Johnson began calling on the radio, "Slate 46, this is Sandy 57."

Jones had been calling on his PRC-90 radio at the top of every hour. None of his calls had been acknowledged, but when he called at 0905, Capt Paul Johnson answered. With voice contact, Johnson could steer to Jones' radio. The bearing he was receiving indicated that Jones was north. Johnson followed the bearing and began to close the distance. As his navigation gear indicated he was close to the survivor, Johnson asked Jones if he could hear an A-10. When Jones replied negatively, Johnson started a descent through the clouds. Eventually, Jones could hear the aircraft and began to give vectors to his location. To aid the process, the pilot expended a bright flare. Jones saw the flare and then the aircraft and directed Johnson right over his position. As he passed overhead, Johnson locked the position into his INS, quickly scanned the area for visual cues, and climbed back above the clouds. Anticipating rescue, Jones described his location to the pilots overhead.

1st Lt Randy Goff and Capt P. J. Johnson, Sandy 57, played a major role in the safe recovery of LT Devon Jones.

All of this was being broadcast on open frequencies.[213] Iraqi forces in the area monitored the conversation and began to react.

Down south at ArAr, Trask and his crew had landed and were refueling with engines running and blades turning. They had been instructed to take off and fly to Al Jouf so that they could get some rest and the aircraft could receive maintenance, but sitting in the refueling pits they monitored the emergency rescue radio frequencies. They heard Johnson talking to the survivor. The AWACS controllers also heard the conversation. Trask again checked with the operations center there and with AWACS. Kingsley with his crew and a fully fueled MH-53 were also at ArAr. The weather was now clear. Colonel Garlington still intended to use the Navy HH-60s from HCS

4/5 on the next CSAR launch. But Trask and his crew had been into Iraq and generally knew the situation. Moccasin 05 was cleared to go. Kingsley was instructed to fly wingman as Moccasin 04.

Trask and crew were airborne moments later with Moccasin 04 on their wing. Northbound, they listened to Captain Johnson talk to Jones. They could not hear his responses, but from what they were hearing, it seemed that Johnson had located the survivor. Unfortunately, as the two helicopters approached the reported location, the Sandys reported that they were once again low on gas and had to depart to find a tanker. Trask was in no mood to delay. "Just give me the survivor's location," he shouted. They were on a non-secure frequency, but Johnson was able to use a secure means to pass the coordinates as determined by his INS. He also gave it to the AWACS, told Jones to come back up on his radio in 30 minutes, and headed south for some gas.

Johnson proceeded back toward Saudi Arabia looking for a tanker. AWACS found another KC-10 for them, but it was 150 miles away. Johnson immediately asked the controller to turn the tanker north since they were low on fuel. The tanker did so and actually rendezvoused with them 30 miles inside Iraq. While the Sandys were refueling, the AWACS began assembling more F-16s and A-10s to support the operation as necessary.

Trask and his crew plotted the location of the survivor as reported by Johnson on their map and loaded the coordinates into their GPS. That gave them a pretty good idea of the overall situation. They realized that the survivor was actually much farther north than they had planned. In fact, he was north of the main east-west highway from Baghdad to Jordan. That line was the de facto dividing line between the rescue forces in Saudi Arabia and the forces in Turkey. Additionally, the road itself was considered to be dangerous. The crew knew what that meant, but they decided to press on.[214]

They used their long-range radio to report their progress to SOCCENT and their intention to cross the road. Trask felt that if they just acted as if they belonged there, nobody would bother them, but he did instruct Kingsley to hold south of the road as they split off and continued north. Crossing the road, they spotted heavy truck traffic, but nobody stopped or appeared to fire at them.

Suddenly, the AWACS reported that a Roland SAM site directly in front of them was active. They plotted the reported location of the dangerous missile site on their map and realized that it was right over the survivor's reported location. Their intelligence study of the Iraqi weapons let the crew know exactly how close they could get to the site before they were in danger, especially at the extreme low altitudes at which they were flying. Trask meticulously kept them on course. They began to get indications on their radar-warning receiver that the Roland site was tracking them. As they pressed on, they spotted the site sitting in the desert. The AWACS controllers began shouting at them to turn to the east. Intent on finding Lieutenant Jones, they pressed in as close as they felt they safely could and then turned east. The whole time the Roland site tracked them but did not fire. They flew off to the east until the Roland was no longer in sight, deciding that the prudent action at this point was to wait for the return of the A-10s.

A few minutes later Captain Johnson and his wingman returned and recontacted the rescue crew. Trask reported that they were in the vicinity of the survivor. "No you are not," Johnson replied. He was back over the survivor's location according to his INS and he could not see the helicopter. Trask assumed that after eight hours of flight, the INS in Johnson's A-10 had probably drifted and asked him to give them a long radio transmission so, they could get a bearing on the location of the A-10s. Johnson did so, and Trask was able to get a bearing that facilitated their rendezvous.

Spotting the helicopter, Johnson and his wingman dropped down to about 1,500 feet and proceeded to fly the standard daisy-chain pattern around them to give what protection they could from enemy guns and SAMs. By now they also had F-15s just above to dissuade any Iraqi pilot from attacking the task force.

Johnson recontacted Jones, who now sensed that rescue was near. The volume of chatter on the radio steadily increased. At one point, as the F-15s started asking for mundane information on the same frequency, Johnson told everybody except

Sandy, Moccasin, and Slate 46 to get off. It worked, and Johnson had only the right players on the frequency.[215]

Then Jones saw the helicopter. He recalled, "For the first time, I looked to the east and saw the Pave Low, about five feet off of the ground. I started talking to him. I have never seen such a beautiful sight as that big brown American H-53 [sic]."[216]

Jones started giving them vectors. Trask was also homing in on his radio signal. Unfortunately, Iraqi forces in the area were homing in on the signal as well. The left gunner on Moccasin 05 spotted two Iraqi army trucks heading directly toward the survivor from the west. Trask called to the A-10s, "We've got movers approaching from our 11 o'clock."

Johnson responded, "What do you want me to do?"

Trask's copilot, Maj Mike Homan replied, "Smoke the trucks! Smoke the trucks!"

The A-10s immediately set up an attack pattern. Johnson rolled in first and just missed the lead truck. His wingman then rolled in and hit the truck with numerous rounds of 30 mm fire. Johnson rolled in again and did the same. The second truck turned around and rapidly left the area.

Trask immediately refocused on the survivor. "Where is he?" he yelled. Johnson replied, "Just land next to the burning truck."

Trask continued inbound. About 500 meters out, he saw Jones stand up. Trask went into a landing flare and set the helicopter down between Jones and the truck, calling to his crew on the intercom that the survivor was at "one o'clock to the airplane and fifty feet." Then he cleared the two PJs to leave the aircraft and get Jones. They were back on board in about 30 seconds. When the PJs reported that they were secure, Trask lifted off and headed south. The A-10s again fell into a daisy-chain above the helicopter for protection.

As Trask headed south, he saw the other MH-53 coming at him. Captain Kingsley had heard all the action and decided to proceed to the area to see if he could help.

Trask was taken by surprise. "What are you doing here?" he shouted.

"I couldn't stand it any more," Kingsley replied and brought his helicopter into formation with Trask.[217]

But they were not yet home safe. They still had to re-cross the main highway. Johnson asked Trask if he wanted to find a quiet spot and try to slip across or if he wanted the A-10s to blast their way across. Trask chose the first option. The Sandys did a quick road reconnaissance, found a quiet spot, and guided the helicopters across uneventfully. Once they were well south and over benign terrain, Johnson split from the rescue birds and went looking, again, for a tanker. He found the same KC-10 that had refueled them the last time and took on another full load of gas. He and his wingman returned to King Fahd Air Base. They had logged over nine hours on the flight—and had the satisfaction of knowing that their efforts had been successful.[218]

Settling down for the long ride home, the crew of Moccasin 05 quickly did some fuel computations and determined that an in-flight refueling might make sense for them, too. They contacted the SOCCENT command center and discussed sending one of the HC- or MC-130 tankers up along the spider route to refuel them. A tanker was launched and orbited along the Saudi border for them, but refueling was not needed. In fact they had enough gas to fly on to Al Jouf. There, Lieutenant Jones was given a quick checkup and returned to his ship. Captain Trask and his crew got some rest and then reported for their next mission.

***

Colonel Comer was elated at the success of his men. They had expected to do their recoveries at night and alone, but this mission showed that they had the tactical flexibility to adapt to a daytime effort as part of a larger task force. But he had other concerns. He sensed that there was a serious flaw or deficiency in their overall capacity to do CSAR. It had to do with locating the survivors and had two dimensions:

1. The bad initial coordinates relayed from AWACS indicated that they could not accurately locate survivors as promised. Comer said, "We thought that if a survivor was transmitting, AWACS or Rivet Joint [RC-135] would be able to give us a coordinate on the survivor within five miles. That is what they were telling us. But, thirty miles was more realistic. Especially from where their orbits were."
2. The navigational devices used by the various coalition aircraft were not necessarily compatible. He added:

We did launch to a coordinate. And the coordinate for Devon Jones was more than 30 miles off. . . . We didn't have the technology on our people who might be survivors to know their location. We had GPS. But GPS was brand new. We didn't understand that it changed the world. That it changed how you fly. That it changed where you fly. We did not know that yet. . . . So we [MH-53s] virtually knew where we were all the time. And that was new, that had never happened before.

These maps, as pretty as they are, were not worth a ding-dong. There are things on this map that are not there. There are things that are obviously there that are not on the map. And things that are on the map are not necessarily correct. These are just estimations because the Iraqis and Saudis do not allow people to go out there and survey to the point of making accurate maps.

Now we can do it by satellite, we can map things accurately. But before this, we had vast expanses of the world that were not mapped. We did not know that coordinates were different. The WGS-84 coordinates that you are getting on the GPS are different if you just read a map. . . . You could be 5 or 8 miles off if you use the wrong kind of coordinate. . . . Those were all things that we didn't know. And GPS was so new that we didn't know that it would change everything. Virtually no one, well the Pave Lows did have it. But nobody else had an integrated navigation system built around GPS. Ours was the first. And the J-model MH-53s were less than a year old. And nobody was in a hurry because this constellation [of navigation satellites] wasn't going to be up until 1993.

But DESERT SHIELD happened and Space Command got involved and . . . started shooting all those satellites up there and then everybody is screaming, "Give me my GPS modification." It was too late for most. That is why the Apaches needed us at the beginning of the war. . . . We were dealing with coordinates from somebody who was flying out there and doing acrobatics dodging missiles with an INS or Doppler, [which is why] his coordinate wasn't going to be close to where that person was on our GPS. We didn't know that. We thought that . . . we should just be able to fly to it and be able to hover above the guy and be able to drop the hoist down through the fog and pick him up. We didn't know.[219]

This was a serious deficiency, because quickly and accurately locating a survivor was fundamental to the CSAR process. In this case, though, the iron determination of the rescue crews in the ad hoc task force prevailed and brought Lieutenant Jones home.

I033 – Ghost 02/AC-130, USAF

The battle against the Scuds continued.[220] Against the advice and wishes of Colonel Gray, CENTAF started using the AC-130 gunships on the mission. Gray felt that the large, lumbering aircraft were much too vulnerable to Iraqi defenses and continued to argue that they be more limited in their use. He was overruled by Glosson and Horner, and the CENTAF/SOCCENT schism deepened.

Working over western Iraq, Ghost 02 with a crew of 13 was engaged sequentially by SA-6 and SA-8 missile batteries. In evading missiles fired by each, the pilot severely overstressed the airframe and engines. He declared an emergency and returned to base. Extensive repairs were necessary. The aircraft did not fly combat again for four weeks.[221]

This confirmed Colonel Gray's fears. He was convinced that the AC-130s were being pushed into areas that were too dangerous. He tried again to regain control over them.[222]

***

In response to the intensity and effectiveness of the aerial operations to date, General Schwarzkopf declared that afternoon that the coalition had attained air superiority over Iraq.[223] Doctrinally, air superiority is defined as, "that degree of dominance in the air battle of one force over another which permits the conduct of operations by the former and its related land, sea, and air forces at a given time and place without prohibitive interference by the opposing force."[224] Obviously, General Schwarzkopf felt that his airmen had seized control of the skies over Kuwait and Iraq and could pretty much do as they pleased.

General Horner was not so sanguine. To him, air superiority was not cut and dried. He wrote,

> Air superiority is not a precise concept. And the process of gaining it is no less fuzzy. What do you mean by air superiority and how do you know when you've got it? There is no handy chart that lets you plot the x- and y-axes and find where the two lines cross. . . . Free operation over Iraq raises other issues. For starters, not every aircraft could be expected to go everywhere. Or if it could go everywhere, it might not be able to go there all the time.[225]

This was how Colonel Gray at SOCCENT felt about his rescue forces. Regardless of the declaration of air superiority by General Schwarzkopf, Colonel Gray, as *the* commander who ultimately had to determine when and when not to send in his helicopters, instinctively knew that there were places in-theater where his recovery crews could not go at a reasonable level of risk.

To further weaken the Iraqi Air Force, the F-111s of the 48th TFW began attacking the hardened shelters on the airfields the next day. They would maintain this subcampaign for the next week, and it would consume 60 percent of the total sorties they flew. The effort ultimately led to the destruction of 63 percent of all of the shelters in Iraq and the aircraft, equipment, and troops inside.

In response, the Iraqi Air Force did two things. First, its pilots began flying their aircraft to Iran, where they were impounded for the duration of the war. Second, they began parking their aircraft next to holy sites, betting that the coalition would not risk damaging the sites in attacks against the aircraft. Both were desperate measures. The practical result was the same—as a fighting force, the Iraqi Air Force was not a factor in the war.[226] While air superiority could be claimed, there were still plenty of SAMs and AAA that had to be dealt with, especially by helicopters.

That evening, the British forces inserted three more SAS teams into western Iraq. Two of the teams, Bravo 10 and 30, were immediately extracted because their drop sites were too dangerous. The third team, Bravo 20, was inserted about 20 miles southeast of Al Qāim. Knowing that the Iraqis expected the coalition commanders to try to rescue their downed aircrews, the team prepared a cover story that they were a CSAR team looking for downed airmen.[227]

Unfortunately, the men of Bravo 20 were spotted the next day by local tribesmen. Unable to contact rescue forces by radio, the eight-man team had to evade. The Brits asked Colonel Comer to launch some MH-53s to recover the men, but while they were planning the mission, it was cancelled. Later one MH-53 and a British CH-47 were launched, but search efforts were unsuccessful. Eventually, three of the men were killed, four were captured, and one man evaded by walking almost 200 miles into Syria.[228]

Concerned for their lost comrades, many of the SAS troops felt the same frustrations as the men of the 4th Wing. The survivor, Chris Ryan, noted that "When the patrol went missing, tremendous pressure built up among the rank and file of the squadron to mount a rescue operation; volunteers were determined to make a box search. When the CO [commanding officer] refused to commit one of his precious few helicopters immediately to the task, some of the guys were on the verge of mutiny. But middle and senior management saw that, in the circumstances, the CO was right to delay the launch until the patrol's situation became clearer."[229]

22 January

That morning Marine major John Steube arrived in Riyadh for duty in the JRCC. He was one of the augmentees requested by Colonel Hampton. A career helicopter pilot, he had recently been the commander of a small search and rescue detachment at Beaufort, South Carolina. After the necessary briefings from the JRCC staff, he made contact with the Marine Tactical Air Control Center and assumed his duties as a Marine liaison officer and duty officer in the JRCC. He felt inadequate since his career had, to this point, been focused on peacetime recovery. He also noted that none of the Air Force personnel in the JRCC had had any specific training in combat rescue. It made him feel a little uneasy. Eventually, the JRCC was also augmented with two Navy officers. Both were career P-3 long-range maritime patrol pilots and had no experience with combat rescue operations.[230]

1035 – Stamford 01/Tornado, RAF

Eight Tornadoes launched a massive attack against the Ar Rutbah radar site in western Iraq with 1,000-pound bombs.[231] Flt Lts Garry Lennox and Kevin Weeks, both on loan from one of the RAF units in Germany, were hit while attempting to loft their bombs. The wingman reported a crash and fireball. No parachutes were sighted, and no beacon or voice was ever heard. Flights working in the area continued to listen for them. Both men had been killed.[232]

Returning to their base, the Brit pilots had to follow a tradition. They went to the rooms of the two lost men and collected all of their personal liquor. Then they took it into the squadron bar and drank it as a toast to the lost men. One of them rationalized later, "If you go to war people are going to be killed. It's one of those facts that you have to accept."[233]

I036 – Magic 11/AV-8, USMC

1st Lt Manuel Rivera stationed aboard the USS *Nassau* was killed when he crashed near Masirah, Oman.[234] A US Navy helicopter recovered his body.[235]

23 January

The F-111s continued to hammer the Iraqi Air Force shelters using laser-guided bombs. They eventually destroyed 141 Iraqi aircraft.[236]

I038/M008 – Wolf 01/F-16, USAF

This aircraft from the 614th TFS was the lead aircraft in a flight of four fragged for an interdiction target near Kuwait City.[237] Immediately after it dropped its bombs, the aircraft burst into flames. The pilot, Maj Jon Ball, was able to glide out over the Persian Gulf before ejecting. The orbiting AWACS monitored his Mayday call and notified the JRCC, who tasked the mission to the US Navy. The Navy RCC launched an SH-60, call sign Spade 50, from HSL-44. It was stationed aboard the USS *Nicholas*, on combat recovery duty in the northern Gulf area. Two USMC AV-8s were diverted to provide escort. Locating the survivor, the helicopter dropped two SEALs who rescued the pilot.[238]

Returning to their base, the pilots in Wolf flight reviewed the mission to determine what shot down Wolf 01. None of the other flight members could recall any active SAM indications or AAA airbursts or tracers. One of the flight members had happened to turn on his HUD (heads-up display) recording device and had filmed Wolf 01 releasing his bombs. The film clearly showed one of its MK 84 2,000-pound bombs detonating just under the aircraft. The fragmentation pattern from the bomb enveloped the aircraft and brought it down.

On that mission, the MK 84 bombs were loaded with special electronic FMU-139 fuses. In investigating the incident, the squadron determined that an "anomaly" had been discovered in the operational testing by the producer of the fuse, Motorola, but any mention of this problem had been excluded from the weapons manuals for the F-16s. The 614th TFS did not use any more FMU-139 fuses in Desert Storm.[239]

The CENTCOM search and rescue plan gave the Navy responsibility for recovery over the Persian Gulf and Red Sea. Allied aircrews were briefed that if they were near the water when hit, they were to go "feet wet" to vastly increase the odds of a successful rescue. The reason was obvious—the US Navy controlled the seas. That procedure facilitated a successful rescue for Major Ball.

In reviewing the significant actions of the last few days, Colonel Gray at SOCCENT felt that he again needed to clarify his requirements for recovery efforts. He sent a message to all units and pointed out that the most significant problem was getting good coordinates on the survivors. He again explained that his helicopter forces had to fly below 100 feet to survive in Iraq and Kuwait. That limited line-of-sight radio contact. Prudently, his helicopters could not loiter for more than a few minutes in the target area. His forces were dedicated to doing everything possible to make a speedy recovery, but the key to the whole process was in quickly getting the best possible coordinates.[240]

The special operations forces guys were more than willing to be the recovery platform. They just could not, in this high-threat arena, provide the "S" in CSAR. Others had to do it for them. And CENTAF controlled most of the assets to do that.

Notes

1. William L. Smallwood, *Warthog: Flying the A-10 in the Gulf War* (Washington, DC: Brassey's Press, 1993), 80.
2. Ibid.
3. Al Santoli, *Leading the Way, How Vietnam Veterans Rebuilt the U.S. Military: An Oral History* (New York: Ballantine Books, 1993), 205.
4. William F. Andrews, US Air Force Academy oral history interview, Colorado Springs, CO, 25 September 1991.
5. Capt Gerry Stophel, interview by the author, 21 February 2002; and 1st Lt Jeff Mase, interview by the author, 9 April 2001. The beer was never delivered. The reason was simple: Gen Buster C. Glosson was given a direct order by GEN Norman Schwarzkopf not to do it. Glosson hears about it to this day. Gen Buster C. Glosson, interview by the author, 25 September 2002.
6. Charles Allen, *Thunder and Lightning: The RAF in the Gulf: Personal Experiences of War* (London, UK: Her Majesty's Stationery Office, 1991), 57.
7. Capt Randy O'Boyle, interview by the author, 20 March 2000.

8. Jay A. Stout, *Hornets over Kuwait* (Annapolis, MD: Naval Institute Press, 1997), 63–65.
9. CPT John Walsh, interview by the author, 6 October 2000.
10. Mase, interview.
11. Glosson, interview.
12. LT Rick Scudder, interview by the author, 29 July 2002.
13. Ibid.
14. LT Jeff Zaun, interview by the author, 17 March 2002.
15. Capt Scott Fitzsimmons, interview by the author, 28 February 2002.
16. Tom Clancy with Gen Chuck Horner, *Every Man a Tiger* (New York: G. P. Putnam's Sons, 1999), 333.
17. Operation Proven Force, "LCDR Speicher Case," http://www.nationalalliance.org/gulf/intel.htm. (n.d.).
18. Stan Morse, ed., "Gulf Air War Debrief, (GAWD)," *World Air Power Journal* (London: Aerospace Publishing, 1991), 118.
19. Lt Col Joe Hampton, interview by the author, 12 March 2000.
20. Michael R. Gordon and Bernard E. Trainor, *The Generals' War: The Inside Story of the Conflict in the Gulf* (New York: Little, Brown and Co., 1995), 200.
21. Ibid., 201.
22. Perry D. Jamieson, *Lucrative Targets, The U.S. Air Force in the Kuwaiti Theater of Operations* (Washington, DC: Air Force History and Museums Program, US Government Printing Office (G. P. O.), 2001), 40.
23. Amy W. Yarsinske, *No One Left Behind, The Lt. Cdr. Michael Scott Speicher Story* (New York: Dutton, 2002), 6.
24. Ibid., 7.
25. Scudder, interview.
26. Yarsinske, *No One Left Behind*, 38.
27. Zaun, interview.
28. Yarsinske, *No One Left Behind*, 38.
29. Sherman Baldwin, *Ironclaw: A Navy Pilot's Gulf War Experience* (New York: William Morrow, 1996), 141.
30. Ibid., 140.
31. Edward Marolda and Robert Schneller Jr., *Shield and Sword* (Washington, DC: Naval Historical Center, Department of the Navy, 1998), 172.
32. Baldwin, *Ironclaw*, 147.
33. Marolda and Schneller, *Shield and Sword*, 167.
34. Gordon and Trainor, *Generals' War*, 122.
35. Gordon and Trainor, *Generals' War*, 207; and LCDR Neil Kinnear, interview by the author, 24 March 2001.
36. History of the Air Force Special Operations Command [hereafter AFSOC History], Hurlburt Field, FL, 1 January 1990–31 December 1991, 86.
37. Kinnear, interview.
38. Rick Atkinson, *Crusade: The Untold Story of the Persian Gulf War* (Boston, MA: Houghton Mifflin, 1993), 32.
39. H. Norman Schwarzkopf with Peter Petre, *It Doesn't Take a Hero* (New York: Bantam Books, 1992), 414.
40. History Channel, *Helicopters*, 1991.

41. Kinnear, interview.
42. Gordon and Trainor, *Generals' War*, 214.
43. Allen, *Thunder and Lightning*, 54.
44. Kinnear, interview.
45. Marolda and Schneller, *Shield and Sword*, 175.
46. Clancy with Horner, *Every Man a Tiger*, 352.
47. Gordon and Trainor, *Generals' War*, 214.
48. Joint Personnel Recovery Agency, Fort Belvoir, VA, USCENTAF/JRCC Incident/Mission Log, DS box, n.d., [hereafter USCENTAF/JRCC Incident/Mission Log].
49. E-mail from Russ Hunter, 26 November 2002.
50. USCENTAF/JRCC Incident/Mission Log.
51. Message, JRCC A/R 180630Z January 1991; Thomas A. Keaney and Eliot A. Cohen, *Gulf War Air Power Survey (GWAPS)* vol. 5, pt. 2, *Summary* (Washington, DC: US G.P.O., 1993), 160; and Hunter, e-mail.
52. Edward Herlick, *Separated by War: An Oral History by Desert Storm Fliers and Their Families* (Blue Ridge Summit, PA: Tab Aero Press, 1994), 33.
53. Yarsinske, *No One Left Behind*, 18.
54. USCENTAF/JRCC Incident/Mission Log.
55. Yarsinske, *No One Left Behind*, 192.
56. Ibid., 30.
57. Ibid.
58. Ibid., 34–40.
59. Center for Naval Analysis (CNA), "Desert Storm Reconstruction Report, Volume II: Strike Warfare" (Arlington, VA: October, 1991), 5-43, 5-57; and GAWD, 226.
60. Gordon and Trainor, *Generals' War*, 222.
61. Ibid., 223.
62. Marolda and Schneller, *Shield and Sword*, 208.
63. AFSOC History, 88.
64. Clancy with Horner, *Every Man a Tiger*, 547.
65. Yarsinske, *No One Left Behind*, 37.
66. Capt Tom Trask, interview by the author, 16 February 2000.
67. Yarsinske, *No One Left Behind*, 41. In 2001 Pres. Bill Clinton changed the status of the pilot, LCDR Michael Speicher, to MIA. See http://www.nationalalliance.org/gulf/intel.htm.
68. Allen, *Thunder and Lightning*, 52.
69. Williamson Murray with Wayne Thompson, *Air War in the Persian Gulf* (Baltimore, MD: Nautical and Aviation Publishing Company of America, 1996), 121.
70. USCENTAF/JRCC Incident/Mission Log.
71. GAWD, 227; and John Peters and John Nichol, *Tornado Down* (London, UK: Penguin Books, 1993), 107.
72. Yarsinske, *No One Left Behind*, 37.
73. Gordon and Trainor, *Generals' War*, 220.
74. USCENTAF/JRCC Incident/Mission Log.
75. Message, JRCC A/R 180630Z January 1991; and *GWAPS*, 160.

76. USCENTAF/JRCC Incident/Mission Log.
77. Operation Proven Force, "LCDR Speicher Case," http://www.nationalalliance.org/gulf/intel.htm. (n.d.).
78. Col George Gray, interview by the author, 3 May 2001.
79. Capt Steve Otto, interview by the author, 30 April 2001.
80. Capt Grant Harden, interview by the author, 2 May 2001.
81. USCENTAF/JRCC Incident/Mission Log.
82. GAWD, 227; and Allen, *Thunder and Lightning*, 57.
83. USCENTAF/JRCC Incident/Mission Log.
84. Hampton, interview; GAWD, 226; and Herlick, *Separated by War*, 98.
85. USCENTAF/JRCC Incident/Mission Log.
86. Robert Dorr, "POWs in Iraq Survived Thanks to Training, Courage, Faith," *Naval Aviation News*, May–June 1991, 6.
87. Zaun, interview.
88. Ibid.
89. Yarsinske, *No One Left Behind*, 98.
90. CNA, "Desert Storm Reconstruction," 5–43; and GAWD, 226.
91. Murray with Thompson, *Air War in the Persian Gulf*, 137.
92. USCENTAF/JRCC Incident/Mission Log.
93. GAWD, 227; and Gray, interview.
94. Gordon and Trainor, *Generals' War*, 229.
95. USCENTAF/JRCC Incident/Mission Log.
96. Marine Corps Historical Center, Washington, DC, 3d Marine Air Wing Command, VMO-2, Squadron history for the period 1 January–28 February 1991, [hereafter VMO-2].
97. Cynthia B. Acree with Lt Col Cliff M. Acree, *The Gulf between Us: Love and Terror in Desert Storm* (Washington, DC: Brassey's Press, 2000), 13.
98. Ibid., 18.
99. Ibid., 20.
100. CNA, "Desert Storm Reconstruction," 5–57.
101. Tom Clancy with Gen Carl Stiner, *Shadow Warriors* (New York: G. P. Putnam's Sons, 2002), 111.
102. Col John Bioty Jr., interview by the author, 6 January 2000.
103. USCENTAF/JRCC Incident/Mission Log.
104. Ibid.
105. LTC Pete Harvell, interview by the author, 29 January 2002.
106. Peter Hunt, *Angles of Attack, an A-6 Intruder Pilot's War* (New York: Ballantine Books, 2002), 237.
107. CNA, "Desert Storm Reconstruction," 5–48, 5–57; and GAWD, 226.
108. Harvell, interview.
109. Marolda and Schneller, *Shield and Sword*, 181.
110. Ibid., 183.
111. Ibid., 194.
112. USCENTAF/JRCC Incident/Mission Log.
113. GAWD, 225.
114. Murray with Thompson, *Air War in the Persian Gulf*, 139.
115. Ibid., 140.

116. Scudder, interview.
117. Michael R. Rip and James M. Hasik, *The Precision Revolution: GPS and the Future of Aerial Warfare* (Annapolis, MD: Naval Institute Press, 2002), 147.
118. Murray with Thompson, *Air War in the Persian Gulf*, 156.
119. Ibid., 150.
120. USCENTAF/JRCC Incident/Mission Log.
121. History of the 614th Tactical Fighter Squadron in Desert Shield/Desert Storm, 29 August 1990–29 March 1991 [hereafter 614 TFS History], Historical Research Agency (HRA), Maxwell AFB, AL.
122. Capt Paul Harmon, interview by author, 16 February 2000.
123. Ibid.
124. From TSgt Ray Cooper, 1723d Special Tactics Squadron: Alert log notes for 19–20 January 1991 (Stroke 65 Mission).
125. Jeff Tice to Paul Harmon, letter, 12 November 1991.
126. Harmon, interview.
127. Lt Col Richard Comer, interview by author, 19 July 2000.
128. USCENTAF/JRCC Incident/Mission Log.
129. Harry M. Roberts, US Air Force Academy oral history interview, 30 August 1991.
130. History, 614 TFS.
131. Murray with Thompson, *Air War in the Persian Gulf*, 161.
132. USCENTAF/JRCC Incident/Mission Log.
133. Ibid.
134. GAWD, 227; Allen, *Thunder and Lightning*, 80; e-mails from Flt Lt Mary Hudson, RAF/AHB, 21, 23, and 27 March 2006; and http://www.raf.mod.uk/gulf/loss.html.
135. Glosson, interview.
136. Murray with Thompson, *Air War in the Persian Gulf*, 164.
137. Schwarzkopf with Petre, *It Doesn't Take a Hero*, 416.
138. Murray with Thompson, *Air War in the Persian Gulf*, 169.
139. USCENTAF/JRCC Incident/Mission Log.
140. Dave Eberly, e-mail to author, 14 April 2002.
141. Rick Atkinson, *Crusade: The Untold Story of the Persian Gulf War* (Boston, MA: Houghton Mifflin, 1993), 126.
142. Ibid.
143. Hampton, interview.
144. Ibid.
145. "Desert Storm, Final Report to Congress: Conduct of the Persian Gulf War, 1991," *Military History Magazine* (Charlottesville, VA: Howell Press, April 1992), 177; and Capt William LeMenager, HQ SOCOM, "A Gulf War Chronicle," unpublished manuscript, January 1998.
146. Gray, interview.
147. Capt Matt Shozda, interview by the author, 12 September 2002.
148. Gordon and Trainor, *Generals' War*, 258.
149. Ibid., 259; O'Boyle, interview; Hampton, interview; and Comer, interview.
150. Gordon and Trainor, *Generals' War*, 258.

151. Time sequence for Corvette 03.
152. Harvell, interview.
153. O'Boyle, interview.
154. Col Ben Orrell, interview by the author, 17 January 2002.
155. Lt Col Bob Ruth, interview by the author, 24 March 2001.
156. Ibid.
157. Lt Col Ed Houle, interview by the author, 28 March 2001.
158. Time sequence for Corvette 03.
159. Ibid.
160. William L. Smallwood, *Strike Eagle: Flying the F-15E in the Gulf War* (Washington, DC: Brassey's Press, 1994), 124.
161. Ibid., 123.
162. Hampton, interview.
163. Smallwood, *Strike Eagle*, 124.
164. O'Boyle, interview.
165. Hampton, interview.
166. Otto, interview.
167. Ibid.
168. Ibid.
169. Benjamin Schemmer, "No USAF Combat Rescue in Gulf; It Took 72 Hours to Launch One Rescue," *Armed Forces International*, July 1991, 37.
170. Gordon and Trainor, *Generals' War*, 259.
171. Hampton, interview.
172. Gordon and Trainor, *Generals' War*, 260; and History of the 4th Wing, HRA (Maxwell AFB, AL: January–December 1991), 77.
173. Shozda, interview; and Lt Col J. D. Clem, e-mail to author, 13 April 2004.
174. Gordon and Trainor, *Generals' War*, 261; Otto, interview; and J. D. Clem, e-mail.
175. Harden, interview.
176. Stovall, interview.
177. Harden, interview.
178. Otto, interview; and Harden, interview.
179. Shozda, interview.
180. Otto, interview; and Steve Otto, e-mail to author, 2 March 2004.
181. Capt Greg Lynch, interview by author, 2 June 2003; and LeMenager, HQ SOCOM, "A Gulf War Chronicle," January 1998.
182. Ruth, interview.
183. Clancy with Horner, *Every Man a Tiger*, 410.
184. Smallwood, *Strike Eagle*, 123.
185. Glosson, interview.
186. Gray, interview.
187. Harvell, interview.
188. Gordon and Trainor, *Generals' War*, 262.
189. GAWD, 194.
190. Murray with Thompson, *Air War in the Persian Gulf*, 172.
191. Ibid., 173.

192. Ibid., 172; and Gray, interview.
193. Gen Sir Peter de la Billière, *Storm Command: A Personal Account of the Gulf War* (London: Harper Collins Publishers, 1992), 191.
194. Clancy with Stiner, *Shadow Warriors*, 436.
195. Yarsinske, *No One Left Behind*, 110.
196. Brig Gen Robert H. Scales, *Certain Victory: The US Army in the Gulf War* (Washington, DC: Office of the Chief of Staff, US Army, 1993), 184.
197. Gordon and Trainor, *Generals' War*, 244.
198. USCENTAF/JRCC Incident/Mission Log.
199. GAWD, 227.
200. Murray with Thompson, *Air War in the Persian Gulf*, 179.
201. USCENTAF/JRCC Incident/Mission Log.
202. Herlick, *Separated by War*, 127; and GAWD, 100.
203. Marolda and Schneller, *Shield and Sword*, 199.
204. Time sequence for CSAR for Slate 46.
205. Detachment Summary Report, HCS-4.
206. Trask, interview.
207. Time sequence for CSAR for Slate 46.
208. Capt Michael P. Vriesenga, *From the Line in the Sand: Accounts of USAF Company Grade Officers in Support of Desert Shield/Desert Storm* (Maxwell AFB, AL: Air University Press, 1994), 178.
209. Trask, interview; and Clancy with Stiner, *Shadow Warriors*, 419.
210. Hampton, interview.
211. 1 SOW History, 1 January–31 December 1991 (Special Operations Wing [SOW] History Office), 130.
212. GAWD, 101.
213. Vriesenga, *From the Line in the Sand*, 180; and GAWD, 98.
214. Trask, interview.
215. Vriesenga, *From the Line in the Sand*, 182.
216. GAWD, 99.
217. Trask, interview.
218. Vriesenga, *From the Line in the Sand*, 183.
219. Comer, interview.
220. USCENTAF/JRCC Incident/Mission Log.
221. *GWAPS*, 169; and AFSOC History, 99.
222. Gray, interview.
223. Clancy with Horner, *Every Man a Tiger*, 548.
224. Joint Staff, Joint Publication 1-02, *Department of Defense Dictionary of Military and Associated Terms* (Washington, DC: 2001).
225. Clancy with Horner, *Every Man a Tiger*, 545.
226. Murray with Thompson, *Air War in the Persian Gulf*, 179–80.
227. Andy McNab, *Bravo Two Zero* (New York: Island Books, 1993), 41.
228. Chris Ryan, *The One That Got Away* (London: Century Press, 1995), 233; and Brig Gen Rich Comer, comments to author on manuscript review, 2 February 2003.
229. Ryan, *The One That Got Away*, 201.
230. Capt John Steube, interview by the author, 9 January 2002.

231. USCENTAF/JRCC Incident/Mission Log.
232. GAWD, 227; and Allen, *Thunder and Lightning*, 80.
233. Allen, *Thunder and Lightning*, 58–59.
234. USCENTAF/JRCC Incident/Mission Log.
235. GAWD, 228; and *GWAPS*, 161.
236. Marolda with Schneller, *Shield and Sword*, 209.
237. USCENTAF/JRCC Incident/Mission Log.
238. GAWD, 227; CNA, 5-58; and USS *Nicholas* (FFG-47), History 1991, Naval Historical Center, Command History (Washington, DC).
239. History, 614 TFS.
240. Gray, interview.

Chapter 5

Desert Storm Weeks Two/Three/Four: 24 January–13 February 1991

We wanted to go up there and obliterate them so badly on the ground that they had no will to fight and no means to do it. We did that. We did that systematically and meticulously.

—Capt Mike Magnus, A-10 pilot

By the second week it was clear that Saddam's forces were being badly mauled. He decided on a new tactic. He directed his commanders in Kuwait to open the spigots in the oil fields and begin pumping crude oil into the Persian Gulf. President Bush labeled this "environmental terrorism." Immediately, General Horner directed his staff to develop a plan for bombing the pumping stations and transfer points. F-111Fs using GBU-15 laser-guided bombs (LGB) stopped the flow.[1]

It was also clear that the coalition forces owned the sky. Saddam's air force was rapidly ceasing to exist, and his air defense units were afraid to even turn on their radars. But attacks on airfields and air defense sites had to continue lest the enemy gain an opportunity to recover from the onslaught.

24 January

I039 – Dover 02/Tornado, RAF

This crew of Flight Lieutenant Burgess and Squadron Leader Ankerson was part of another large force of Tornadoes again attacking the Ar Rumaylah Airfield in Iraq.[2] They were forced to eject when one of their 1,000-pound bombs detonated prematurely and destroyed their aircraft. Ankerson was met by Iraqi troops and immediately captured. Burgess was able to slip away in the dark and hide behind a small ridge. He activated his beacon and made several radio calls. None were acknowledged. At dawn, he could see that he was in an area of bunkers and

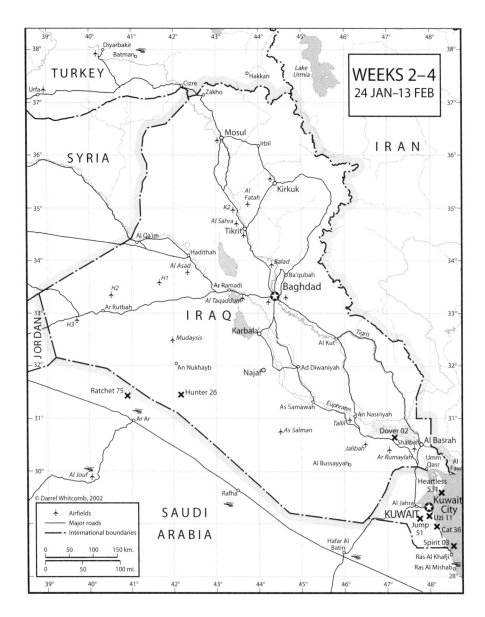

enemy troops. An estimated two hours later, a jeep of soldiers drove directly up to him and took him prisoner.³

This was the fifth Tornado loss in seven days and the second regrettable occurrence of fratricide. The airfield attacks with the Tornado/JP.233 combination had been critical to shutting down the Iraqi Air Force and attaining air superiority, but now that had been done, and it was time to reevaluate tactics. That reevaluation clearly indicated that low-level attacks were no longer necessary. The RAF units joined the medium-altitude war. As one of the commanders recalled, "It was a straightforward operational decision that at that stage we didn't need to go down and drop weapons at low level. It made more sense to bring our aircraft up, so we changed tactics and got into the precision-guided business."[4]

Accordingly, the British Tornadoes were retasked for attacks on bridges. The RAF reinforced its Tornado units with Buccaneers equipped with laser designators, and they teamed up with the Tornadoes for precision strikes. Using LGBs, the Brits destroyed 63 percent of the bridges they attacked.[5]

This, too, would carry its own risks because the British had not made the investment in electronic jamming pods for their aircraft, as the Americans had. This meant that if they encountered a radar-controlled SAM and had no Wild Weasel support, they would be vulnerable. But that was still better than being down among the guns.

I040 – Active 304/F-18, USN

Returning to the USS *Roosevelt*, the Navy F/A-18 experienced engine failure over the central Persian Gulf.[6] The pilot radioed his predicament to the Marine TACC and ejected. He was injured in the process. Orbiting aircraft established contact with him and alerted the Navy RCC. They dispatched an SH-3, call sign Venom 505, from the USS *Caron* with two SEALs on board for the recovery. The pilot was picked up and taken to the USS *Wisconsin* for medical treatment.[7] As with Wolf 01 the day before, the aircrews knew that there was relative safety in ejecting over the coalition-controlled Gulf.

25 January

I042 – Desert Hawk 102/UH-60, RSAF

On a support mission, the helicopter crashed 65 miles northeast of Al Jabayl. The Army RCC monitored its emergency call and responded to recover the crew. This was a classic SAR situation. In this case, US assets recovered a Saudi crew.[8]

26 January

As the air campaign progressed, the coalition ground combat units were training and repositioning for the coming ground campaign. All US Army and Marine units had large fleets of helicopters. The Army RCC and the Marine TACC were the immediate points of contact for any SAR or CSAR efforts regarding their aircraft.

1045 – Bulldog /OH-58, US Army

The aircraft crashed in a sandstorm near Al Qaysūmah, Saudi Arabia.[9] The Army RCC responded and sent a recovery force to safely recover the crew.[10]

***

Early that afternoon the CENTAF commander, General Horner, sat down and reviewed the first nine days of combat. Looking back at the effort expended thus far, he felt that the objectives of Phase I and II of the air campaign had mostly been met. He roughly calculated how many sorties would be needed over the next several days to complete the campaign and sustain the damage. He then turned to the next phase, the preparation of the battlefield. That would require a refocusing of the air effort.[11]

Horner commented to the CENTAF staff, "We are where we need to be to shift the emphasis to the Republican Guards."[12] This fit in with what his boss wanted. General Schwarzkopf was thinking ahead to the ground battle. He wanted the air effort to refocus on the Iraqi army, and specifically, the Republican Guard divisions.

To facilitate the flow of aircraft over a relatively small area, the area between the Saudi-Iraqi-Kuwaiti border and Basra and east of Hafar al Batin was divided into kill boxes. Each 900-square-mile box was defined by specific latitudes and longitudes. Different forward air controllers or flights would be assigned to work in each. This was an effective deconfliction measure.[13]

To further refine the effort, General Horner had the planners focus the attacks each day on a specific division. One day would be "Hammarabi day" or "Al Fāw day." Eventually more than 750 sorties a day were directed against the identified ground units.

27 January

The next day Colonel Comer at ArAr was notified that one of the British SAS teams inserted into western Iraq a few days earlier was missing. The team, Bravo 20, had last reported its position as about 50 miles southeast of Al Qāim. Comer was directed to begin planning a rescue mission for them. The unit intelligence section had done an excellent job of templating the enemy forces and defensive systems in the area. This allowed the crews to develop a comprehensive plan for searching for the team while paying due respect to the dangerous enemy SAM sites and AAA guns. This was a special operations mission, and no incident or mission number was assigned.

When ordered to launch, an MH-53 of the 20th SOS, piloted by Capt Corby Martin and Capt William LeMenager, took off in formation with a British CH-47 Chinook from the RAF 7 Squadron. The crews proceeded to the team's initial insertion point and flew their preplanned evasion route back toward the Saudi border. No contact was made with the team. Using the intelligence data given them, the crews then began an intensive search of the area. At one point the ever-present AWACS called them with a warning of an active SA-8 missile site directly in front of them. They altered course and ended up flying very close to the Mudaysis Air Base. When their fuel was exhausted, they began to head back to ArAr.

Unexpectedly, the crew of the CH-47 began calling for Bravo 20 on the emergency radio frequency. This had not been briefed to the US crew and took them quite by surprise. A few minutes later they were shocked to see the Chinook crew dispensing flares. This illuminated the two helicopters, and LeMenager called them to find out what they were doing. The Brits replied that they were trying to gain the attention of the ground team.

A few miles north of the Saudi border, the radar-warning receiver on the MH-53 picked up a signal from another SA-8 site. They took evasive action and dispensed chaff to fool it.

The search for Bravo 20 was unsuccessful, but it showed that the Pave Lows could be used for search missions where the enemy threat was more permissive—especially at night. The key was good mission preparation and preplanning.[14]

That same evening at a press conference, General Schwarzkopf declared that coalition air forces had achieved air supremacy over the Iraqi nation. That meant that coalition aircraft could range over Iraq with impunity and do whatever they wanted to do.[15]

That assessment applied primarily to the fast mover aircraft that could operate above 10,000 feet. To the helicopter pilots, it did not mean as much. But in analyzing the missions flown to date, Colonel Comer felt that at night, his helicopters could safely fly over as much as 70 percent of Iraq at a reasonable level of risk, as they had on the Bravo 20 mission. But, they still had to avoid the cities and known troop concentrations, and there was still the unpredictable danger of flying over roving bands of unfriendly Bedouins.[16]

28 January

I047/M009 – Cat 36/AV-8, USMC

Capt Michael Berryman launched as the flight lead of two Harriers on ground alert in response to tasking from the Marine ground units.[17] They attacked a FROG missile battery on the Kuwaiti coast. Enemy AAA was heavy in the area, and the ground forces had heat-seeking missiles. Berryman's shootdown was not observed, and he was reported missing when he failed to rejoin his wingman after the air strike.[18] Another flight of AV-8s and an F-18 searched for him without contact. Intelligence sources subsequently determined that his aircraft had been hit by a heat-seeking missile, and Iraqi forces had immediately taken him prisoner.[19]

I048/M010 – Jaguar, Omani Air Force

This aircraft was reported missing 12 miles south of Dhahban, Oman.[20] A SAR effort was initiated. Local forces, USAF C-130s, and an RAF Nimrod searched. The pilot was the station commander. He was found dead by Omani rescue personnel and classified as a noncombat loss.[21]

***

Intelligence sources indicated that the focused attacks on the Iraqi forces in Kuwait were beginning to take a toll. Saddam and his military leaders appeared to have believed that any ground attack would be preceded by no more than three to seven days of focused air attacks specifically on their forces. They were shocked at the relentlessness of the campaign. They appeared equally shocked when this newly focused air attack did not seem to be a prelude to the ground attack either but a sustained attack in its own right. They were also alarmed at the effectiveness of the shelter-busting attacks by the F-111Fs and the lack of desired effects from their Scud attacks against Israel and Saudi Arabia.

Perhaps desperate, the Iraqis decided to launch a "spoiling attack" in Kuwait with three divisions: the 1st Mechanized, the 5th Mechanized, and the 3d Armored. It would be supported by commandos coming south along the coast in small boats. Such an attack would have two objectives: first, to cause American casualties that would stir up political opposition in the United States; and second, to provoke the coalition into initiating a ground war on the Iraqis' terms. This would turn the war to what was their primary strength, their powerful ground units.[22]

Having been provided US intelligence during the Iran-Iraq war, the Iraqis knew the times when US reconnaissance satellites were overhead. They timed their moves so that they had the least chance of detection. That evening a brigade-size task force from Iraq's 3d Armored Division began moving toward the Saudi border. Moving quickly, its lead units were almost to the border before US Marine outposts spotted them. The marines called in air support and began to decimate the attacking elements.[23]

The Iraqis persevered and slammed into several outposts. They had studied the operations of the Marine Corps and understood their procedure for calling in air strikes. At one point, they used powerful radio jammers to block radio frequencies and keep the marines from calling for air support. The Marine outposts used backup telephone lines and were able to get some air sorties to the initial engagements. Unfortunately, in the confusion of the individual battles, there were two incidents of fratricide, and several marines were killed before the attacks of the 1st Mechanized Division and the 3d Armored Division were blunted.

30 January

The attack of the 5th Mechanized Division came down the main coastal highway and entered the town of Khafji, five miles inside Saudi Arabia. Coalition leaders were slow in picking up the significance of the Iraqi move. Khafji had been abandoned because it was within range of Iraqi artillery, but once the penetration was discovered and Horner became aware of the situation, he ordered the TACC to begin flowing air to the battle. Additionally, he directed some special reconnaissance assets to reposition to better observe the enemy. Aircraft of all types were sent against the invaders.[24]

The small boats carrying the commandos were also spotted and attacked by A-6s with precision bombs and by Royal Navy attack helicopters from the British warships HMS *Brazen*, *Cardiff*, and *Gloucester*. Firing 25 Sea Skua missiles, they damaged all 11 of the Iraqi boats and drove them off.[25]

That evening a task force of Saudi mechanized infantry and Qatari armor attacked the invaders in Khafji. They were beaten back after two uncoordinated attacks. US air strikes pounded the Iraqis in and around Khafji. Other air strikes destroyed truck columns that tried to resupply or reinforce the invading units. The next morning, an AC-130 gunship was refragged to attack the two Iraqi divisions.

31 January

I051/M011 – Spirit 03/AC-130, USAF

With its massive firepower, the AC-130 was a tremendous asset in any land battle. It had more than proven its worth 20 years prior in the skies over Southeast Asia. But the big, lumbering aircraft was also vulnerable to ground threats. The best tactic was to work at night and hide in the darkness.

Accordingly, Spirit 03 was launched into the night to support the US Marine and Saudi ground forces containing and counterattacking the Iraqi forces of the 3d Armored Division that attacked in Saudi Arabia just north of the town of Khafji.[26] Finding many targets, the Spirit 03 crew provided excellent support to the coalition ground units below. As dawn approached, one of the controllers aboard the sector AWACS contacted the Spirit pilot and warned him of pending dawn. "I can't go right now," aircraft commander Maj Paul Weaver responded. His crew was engaged in destroying a FROG missile battery. The FROG missiles were a serious threat to coalition ground forces in the area. Less than a minute later, the aircraft was hit by an enemy missile. The deadly projectile struck one of the engines on the left wing. Somebody on the aircraft called "Mayday" on the radio. That call was the crew's last transmission. The explosion caused the wing to crack, and the aircraft plunged into the water just a few hundred yards off the coast. As soon as the loss was reported to the JRCC, SOCCENT dispatched aircraft into the area to search for possible survivors. The 55th SOS was alerted for recovery. The wreckage was observed falling into the water, and efforts focused there. Helicopters from the USS *Nicholas* and several Saudi boats searched the area, but there were no survivors.[27] Killed were:

Maj Paul Weaver	TSgt Robert Hodges
Capt Arthur Galvan	SSgt John Blessinger
Capt William Grimm	SSgt Tim Harrison
Capt Dixon Walters	SSgt John Oelschlager
1st Lt Thomas Bland	SSgt Mark Schmauss
MSgt Paul Buegge	Sgt Barry Clark
MSgt James May	Sgt Damon Kanuha[28]

Photo courtesy of Peter Mersky

MH-53s stood CSAR alert near the Saudi-Iraqi border at ArAr.

As ground forces held the penetration and a renewed Saudi attack swung around to the north, cutting off the Iraqi unit in the town, airpower mauled the invaders. One tank brigade was virtually destroyed when it got caught in a minefield just as a number of flights of A-10s, AV-8s, F-15Es, and others decimated it with bombs and CBUs. A survivor of this attack was later captured. He stated that his unit took more damage in that one attack than it did in 10 years of war against Iran. The air strikes against the 3d Armored Division were so unremitting that the unit never had a chance to concentrate and mount a focused attack.[29]

Sensing the moment, the Saudi commander again ordered his Saudi-Qatari task force to attack and seize the town. This time, the Iraqis broke. Out came the white flags, and the task force secured its objective.[30]

The battle of Khafji was short but fierce. It intensified the focus of CENTAF on the Iraqi forces in Kuwait or just north of it. To preclude a repeat of the attack, the forces along the border received the largest share of the bombs, but no Iraqi unit or facility south of Basra was safe, and the air campaign against them was relentless.

As the battle of Khafji was being fought, elements of an 877-man US joint special operations task force (JSOTF) began to deplane at ArAr. Arriving directly from SOCOM-assigned units in the United States, they began to acclimate and train for the upcoming operations. Their commanders spent time with the commanders and crews of the MH-53s, HH-60s, MH-60s, and the A-10s, as they prepared to fly into western Iraq to reinforce the British SAS in its campaign against the Scud sites and missiles. This was under the direct command of CENTCOM.[31]

1 February

The next morning Iraqi naval units in the northern Gulf attempted to make a dash for safety in Iran. Coalition naval elements intercepted them. For more than 13 hours, coalition ships, helicopters, and attack aircraft pummeled the fleeing armada. Coalition aircraft destroyed or damaged seven missile-carrying boats, three amphibious ships, a minesweeper, and nine other small craft in the shallow waters near Bubiyan Island. US Navy combat recovery forces were kept busy pulling surrendering Iraqi sailors and soldiers out of the water. Only one ship made it to Iran and was impounded for the duration of the war.[32]

2 February

Now two weeks into the war, CENTAF operations officers did a quick analysis of CSAR operations to date. To reemphasize to the aircrews the need for tactical flexibility in CSAR operations, the director of operations at CENTAF sent this message to all units:

> Subj: Operation DESERT STORM Evasion and Escape Tips. (U)
>
> (U) Combat search and rescue recovery may be executed by any of the following: helicopters with or without protective fighter aircraft; naval vessels, with possible air cover; armored vehicle reconnaissance; infantry units as part of an advance. All of these methods may involve a surprise move by the recovery force to at least temporarily overwhelm the enemy with superior firepower in the vicinity of the evader(s). Such efforts usually require speed to prevent the enemy from increasing his strength. Evaders that can move may be able to improve their chances of being successfully recovered by conducting initial evasion travel to a suitable hole-up site, employing discreet communication and signaling procedures, and selecting a site for recovery that considers the potential enemy opposition in the area.[33]

As this message was being sent, the intense battles in the northern Gulf area continued unabated. Naval strike packages routinely flew sea suppression missions in the area, hunting for Iraqi naval forces. The packages usually consisted of two A-6s armed with bombs and EA-6Bs, which flew above and used their powerful electronics to jam Iraqi radars. The EA-6Bs also carried HARMs that could home in on and destroy any radar units foolish enough to radiate long enough for the missiles to find them.[34]

I052 – Heartless 531/A-6, USN

This aircraft from VA-36 off the USS *Roosevelt* launched with a crew of LT Pat Connor and LCDR Barry Cooke to fly in one of the armed surface reconnaissance missions east of Faylaka Island in the northern portion of the Persian Gulf.[35] The crew reported that they had been fired at with an SA-7 heat-seeking missile. A few minutes later they were observed pursuing a patrol boat. The crew did not answer subsequent radio calls. A visual search of the area by other crews in their flight and a Navy SH-60 revealed an oil slick and drop tank floating on the water. Enemy SAMs and AAA were active in the area and were engaged by numerous aircraft. Intelligence sources subsequently determined that the crew had been killed in the engagement.[36] Connor's body washed up on one of the offshore islands at the end of March.[37]

I053/M012 – Uzi 11/A-10, USAF

Not far to the west, coalition aircraft continued to strike enemy units in the kill boxes between the Saudi-Kuwaiti border and Basra in Iraq. Two A-10s led by Capt Dale Storr struck targets in the AG4 kill box in Kuwait.[38] Lacking a forward air controller, they were assigned a target just north of the Al Jaber Airfield by the local airborne controller. They initially dropped some bombs and fired their Maverick missiles on a warehouse complex. The enemy gunners responded and began to engage with mixed-caliber AAA. Captain Storr then spotted a convoy of enemy trucks. He made two strafe passes and hit several. As he was recovering from his second pass, his wingman, 1st Lt Eric Miller saw what appeared to be a brown puff of smoke beneath his leader's aircraft. Then the stricken plane rolled off to the right and began an out-of-control descent. Storr fought to control the aircraft, but the flight controls had been fatally damaged. Miller called his flight leader repeatedly, but Storr did not respond. Desperate, he called for Storr to eject. Storr heard the call and did so. Miller did not see Storr or his parachute. Instead, he saw the aircraft hit the ground and explode.

Storr had ejected below 2,000 feet. As his chute opened, he was immediately overcome by the harsh sound of the war raging below him. He could hear enemy bullets whizzing by. His parachute carried him through the smoke of his burning aircraft.

Landing unhurt on a small sand dune, Storr grabbed his emergency equipment and immediately tried to find some cover. He spotted enemy troops approaching in a truck from the general area that he had just strafed. Storr had two survival radios. The one in his vest was a PRC-90. He had another in his emergency bag. Unfortunately, it was for training purposes only and was not tuned to the appropriate frequencies. That was the one he grabbed. He made several desperate calls to Miller that his wingman did not hear because he was monitoring another frequency. The enemy troops closed in and captured him. Some struck him with their fists and one with a rifle, but otherwise, he was unhurt.

Unaware of any of this, Miller set up an orbit and began calling for his flight lead. He also notified the orbiting AWACS that Uzi 11 was down. Immediately a forward air controller, Nail 55, Capt Mike Beard, joined him over Storr's location. They repeatedly called on the emergency frequency but never heard a response. Miller spotted several enemy trucks rapidly driving into the area and strafed them.

By now Miller was low on fuel and departed for a tanker, as Nail 55 continued to orbit over the downed aircraft. Two A-10s on rescue alert at KKMC were scrambled. They arrived over the location and assumed the on-scene command.

Miller returned after refueling, but there was no contact with Storr. When his fuel was exhausted, Miller returned to KKMC, where he met with SOCCENT personnel. They questioned him closely about the crash site and lack of contact either electronically or visually with Storr.[39] A package of rescue helicopters from the 55th SOS and the 71st SOS was launched and flew near the Kuwaiti border. Colonel Hampton worked closely with the battlefield coordination element (BCE) in the TACC to shut down US artillery so the helicopters could safely pass.[40] Lacking any contact with the survivor, and considering the enemy threat in the area, they were never committed into Kuwaiti airspace. The next day intelligence sources reported that Storr had been immediately captured by Iraqi ground forces.[41]

As the aerial forces were pounding the Iraqis, various US and coalition ground units continued to prepare for the pending ground war. Their helicopters were constantly in action.

I054/M013 – Millcreek 701/AH-1J, USMC

Maj Eugene McCarthy and Capt Jon Edwards were lost while flying with NVGs as escort for a medical evacuation mission out of the Kibrit Airfield in northern Saudi Arabia.[42] When they did not return to their base, an immediate search was initiated. A British Nimrod, Marine OV-10s, and AH-1s searched without success. The next morning a Marine ground team found the crash and recovered the remains of the two men. This was subsequently determined to be a noncombat loss.[43]

I055 – Hulk 46/B-52, USAF

B-52s were heavily involved in the war. None were based in the region. Instead, they flew in from the United States, England, or Spain. The United States was also granted permission by the British to use their base at Diego Garcia in the Indian Ocean. This meant that all B-52 missions were long flights, which strained these already old aircraft. This was their second major war, and the aircraft suffered continuous mechanical problems.

Hulk 46 suffered a massive electrical failure while lowering its flaps for landing at Diego Garcia, causing the fuel pumps to shut down.[44] Loss of positive fuel pressure precipitated the flame-out of all engines. When the aircraft went down, another B-52 and two KC-135s in the area diverted to set up a search pattern. A US Navy SH-3 helicopter immediately responded and picked up three crewmembers who had safely ejected. The other three men on board were killed.[45]

***

To increase the effectiveness of his tactical strikes, General Glosson decided to use fighter pilots in F-16s to direct air strikes. Called "killer scouts" to differentiate from forward air controllers in A-10s and OA-10s, these pilots were drawn from the 4th TFS, led by Lt Col Mark Welsh. They would be assigned to the individual kill boxes in the KTO, which had been established earlier. A flight of two scouts would be assigned to each box, as the ground situation dictated, to find enemy targets. The scouts would then direct other flights of fighters, bombers, or even gunships as they hit the targets. The scouts could also accurately determine the effectiveness of the strike to help better assess how the air campaign was progressing.[46]

I056 – Gunfighter 126/UH-1N, USMC

On a routine mission to resupply forward units, this aircraft from Marine Light Attack Helicopter Squadron 369 crashed for unknown reasons.[47] All bodies were recovered. This was determined to be a noncombat loss.[48]

5 February

I061 – Warparty 01/F-18, USN

While returning to his ship, Warparty 02 reported that his wingman, LT Robert Dwyer, was having navigation trouble and that the flight had become separated.[49] The Navy RCC began a search for the aircraft. Radar controllers on the USS *Bunker Hill* reported that they had tracked an aircraft that descended from 30,000 feet to the surface in 60 seconds. From their radar tapes, they determined the missing aircraft's last known position to be 25 miles southeast of Faylakah Island in the northern Persian Gulf. Rescue helicopters were ordered into the area. Fighter aircraft provided security while they searched. The helicopter crews discovered an oil slick but no other indications of the aircraft or pilot. After two days, the search was terminated.[50]

***

The pummeling of the Iraqi forces continued. Some F-111s were thrown into the mix. Working in the kill boxes over Kuwait, some of the crews noticed that their heat-sensitive forward-looking infrared sensors (FLIR) could detect Iraqi tanks and other pieces of heavy equipment like artillery and AAA guns. The commander of the F-111F wing, Col Tom Lennon, led a flight of two aircraft carrying 500-pound LGBs. Using their FLIRs, they attacked eight Iraqi vehicles in the Medina Division, a prime target. Analyzing the film, Colonel Lennon concluded that they had destroyed seven of the targets. The F-111s were quickly rearmed and refragged to join the attack against the Iraqi forces. Their priorities were chemical weapons launchers, artillery, tanks, and other armored vehicles. Lumped together under the moniker of "tank plinking," the effort added to the systematic destruction of the Iraqi ground forces.[51]

Within days, General Glosson ordered the F-15Es and A-6Ds, with their laser bomb capability, to join this attack. He also ordered the F-16s—the most common aircraft in the theater—to press their attacks below 9,000 feet so that they could be more accurate and increased the missions for the F-16s flying as killer scouts.[52] This was a throwback to the old "fast forward air controller" program in Southeast Asia where fighter crews in F-100s or F-4s did the same thing in high-threat areas.

To protect the attack aircraft, flights of F-4G Wild Weasels would orbit overhead and launch HARMs at any SAM or other radar-controlled sites that attempted to track coalition aircraft. Whenever they launched a missile, the Weasels would call "Magnum" on the radio to alert other crews that a HARM had been fired. The other fighter crews noticed how the enemy radar sites would turn off their radars when "Magnum" was called. They adopted the tactic, and whenever Weasels were not around, they would call "Magnum" anytime a radar site went active.[53] The overall result of these tactical changes was a more focused and intense air campaign against the Iraqi forces in and around Kuwait.

6 February

I063 – UH-1 Medical Evacuation, US Army

This helicopter crashed 20 miles east of Rafha.[54] The Army RCC responded and dispatched a recovery aircraft (Dustoff 229). One crewmember was killed, and four were wounded. This was a noncombat loss.[55]

***

The level of effort against the enemy ground units continued to grow. Fifty more sorties a day had been added to the attack force hitting the Iraqis in Kuwait. Most of the increase was the F-111s, which continued to plink the tanks and artillery. The planners began to move the strikes deeper into Kuwait. Marine air strikes focused on enemy units across the border from its ground units. The Navy continued to pound targets along the coast. Even SOCCENT got into the attack by sending one of its MC-130 Talons to drop a 15,000-pound bomb on Iraqi troops. The strike was designed to lower their morale. Intelligence data indicated that it did.[56]

7 February

In the early morning hours helicopters from SOCCENT began to insert elements of the JSOTF into western Iraq to reinforce the British SAS in its campaign against the Scuds. Coordinating with the SAS, they operated in the area between Al Qāim and the airfields to the south called H-2 and H-3. Over the next weeks, they continuously called in air strikes on the enemy forces. In one action, a reinforced platoon of US Army Rangers attacked and destroyed a large communications facility near the Jordanian border. Working with F-15Es that attacked nearby AAA sites, they also used explosive devices to drop a 350-foot communications tower.[57]

Special Operations MH-60s and MH-6s from the 1st/160th SOAR (1st Battalion of the 160th Special Operations Aviation Regiment) began flying "thunder run" missions in direct support of the now marauding teams throughout the area of operations. NVGs made them especially effective at night.[58]

The efforts of the British SAS and the US JSOTF paid dramatic dividends. Since the Scud launches began 18 January, the Iraqis had launched 29 missiles from the western desert. After 7 February, they only launched 11 more, and two of those fell harmlessly into the desert.[59]

Flying support for the operation kept the helicopter crews of SOCCENT busy. Still, at no time did it keep them from having helicopters on alert at several locations for combat recovery.[60]

The dual tasking was at times confusing to the aircrews. Capt Paul Harmon explained that often crews on rescue alert were kept busy planning for SOF support missions. "One of the things that I think is important—even though we were tasked with other missions, we never came off alert," he noted. "We might have been pulling CSAR alert when we were doing planning cells. That's one of the gross misconceptions over the last 10 years, [that we] were too busy doing these other things, that we didn't have time to do CSAR. [It's] not the case."[61]

Subsequently, he and the other Pave Low crews flew support missions for the JSOTF operations in western Iraq but were always on alert for combat recovery. In some cases, this meant that crews had to violate crew rest rules. When faced with complaints, 20th SOS commander Colonel Comer granted "squadron internal waivers," knowing full well that what he was doing was clearly illegal. But war was war.[62]

9 February

I064/M014 – Jump 57/AV-8, USMC

On his 17th combat mission, Capt Russell Sanborn was hit by what appeared to be a heat-seeking missile as his flight of two AV-8s attacked an artillery battery in Kuwait.[63] He was the wingman in the flight. His flight lead was the Marine Air Group (MAG) 13 commander, Col John Bioty.

When the missile slammed into the AV-8, the aircraft flipped over and entered a spiral. Realizing that the aircraft was no longer flyable, Sanborn ejected. Upon landing he quickly spread out his orange and white canopy so aircraft above could spot his position. Unfortunately, he had landed near an enemy headquarters unit, and Iraqi soldiers quickly closed in on him. Sanborn checked that his pistol was cocked and then made a quick call on his PRC-90 survival radio that he was down and alive. The call was not acknowledged. Shouting at him in Arabic, the enemy troops closed in, and he was immediately captured.

Nobody reported hearing either Sanborn's emergency beacon or voice transmission. The Marine TACC reported the event to the JRCC, who tasked SOCCENT. SOCCENT launched two MH-60s from the 55th SOS and an MH-47 from the 3d/160th. They held short of the front lines pending contact with Sanborn. They were escorted by two A-10s.

Sanborn's parachute was sighted, but intelligence sources relayed reports from Kuwaiti agents that Sanborn had been captured by Iraqi troops and was slightly injured. Regardless, the A-10s searched for the pilot until they had to leave for fuel.

As a result of this incident, AV-8s began carrying and using decoy flares to protect them from the dangerous enemy missiles.[64]

11 February

General Powell and Secretary of Defense Cheney flew to Riyadh for a high-level review of the war. After they were briefed on results to date, the discussion turned to ground operations. Each senior commander was allowed to give his positions on the war. Secretary Cheney then set the date for initiation of the ground war as 21 February or a few days after, depending on the weather.[65]

13 February

As the strategic campaign continued, F-117s were retargeted against command and control centers. One command center in particular, near Baghdad, was noted for its high level of activity. In the early morning hours, the Al Firdos bunker was hit with two 2,000-pound LGBs. It appeared to be a successful routine strike.

I069/M015 – Hunter 26/F-5, RSAF

Although the US Air Force, Navy, and Marines flew the majority of sorties in Desert Storm, our allies also made significant contributions to the war. The British and Italians had already lost several Tornadoes in the dangerous airfield attacks. All forces suffered battle damage.

The Saudis reported that Hunter 26 had gone down just inside Iraq and north of ArAr.[66] Two HH-60Hs from HCS-4 at ArAr launched and flew search patterns with A-10 escort. Voice contact and authentication with the pilot were never received, and the aircraft were recalled. Search efforts continued for several days with RSAF UH-1Ns and C-130s and a US Navy E-2.[67] The pilot had been captured by the Iraqis.

I070 – Tiger 53/F-15, RSAF

This aircraft from the 6th Squadron crashed 25 miles southeast of Mushait Air Base.[68] This mission was controlled by the RSAF. Their helicopters launched, found the wreckage, and reported that they had recovered the body.[69]

I071 – Ratchet 75/EF-111A, USAF

This late in the war Iraqi aircraft almost never ventured into the sky. But that did not mean that they were not still a threat. Ratchet 75 was on a mission in Iraq to jam enemy radars.[70] It was the third aircraft in a flight of four. Shortly after it reached its orbit point in Iraq, two F-15s passing south through the area saw an aircraft below them eject a series of countermeasure flares. Twenty seconds later they saw the same aircraft eject three more flares and then disappear in a huge ball of flame as it hit the ground just inside the Saudi border near the town of ArAr.[71] Killed were the pilot, Capt Douglas Bradt, and the electronics warfare officer, Capt Paul Eichenlaub.[72]

The wreckage was sighted from the ArAr tower. HCS-4 crews planned a mission to the site but were never given launch authority by SOCCENT. Instead, aircrews from the 3d/160th easily found the crash site and the dead crewmen. HCS-4 helicopters then flew a SOF/USAF security team to the site and supported them for the next several days as they searched the site and recovered the remains of the crew.[73] Although listed as a combat loss, the crash was blamed on aggressive defensive combat maneuvering of the crew.

***

EF-111s had been attacked on 19 January by a MiG-25, which launched three missiles at their formation. All successfully evaded the attack, but the MiG effectively broke up their jamming pattern and continued on to shoot down Corvette 03. The EF-111 crews also feared the ever-present F-15s because they were aware of at least one instance when an F-15 had been cleared to fire on an EF-111 by an AWACS controller. The pilot held fire when he closed to within visual distance and confirmed that the aircraft was a friendly.[74]

These events put real fear into the hearts of the EF-111 crews. Several subsequently reported that they were being intercepted by Iraqi MiGs while in their orbits, but the attacks were never confirmed. Regardless, among the EF-111 crews, their suspicions evolved into the legend of "Baghdad Billy," the nighttime MiG pilot who went looking for lumbering EF-111s.

Later speculation suggested that the sensitive radar-detection systems on the EF-111s were actually picking up radar signals from the F-15s and presenting them as possible MiGs trying to find and shoot them down. Perhaps the crew of Ratchet 75 had picked up such a signal and taken evasive maneuvers to avoid what the crew thought was a MiG. In trying to evade, they flew the aircraft into the ground—the ever-present killer.[75]

The EF-111 pilots wrote a song about "Baghdad Billy":

I'm an F-111 Jock, and I'm here to tell
Of Baghdad Billy and his jet from Hell.
We were well protected with Eagles in tight
But that didn't stop the man with the light.

RJ, AWACS—they didn't see
As Baghdad Billy snuck up on me.
Then I found a spotlight shining at my six
And my woozoo said "hooooly . . . ."

I popped off some chaff and I popped a flare
But that Iraqi bandit, he didn't care.
I had tracers on my left, and tracers on my right.
With a load of bombs, I had to run from the fight.

I rolled my "Vark" and took her down
Into the darkness and finally lost the clown.
When I landed back at Taif and gave this rap,
CENTAF said I was full of crap.

I am here to tell you the God's honest truth.
That Iraqi bandit, he ain't no spoof.

You don't have to worry, there is no way
You'll see Baghdad Billy if you fly in the day.
But listen to me son, for I am right,
Watch out for Baghdad Billy if you fly at night.[76]

Notes

1. Williamson Murray with Wayne Thompson, *Air War in the Persian Gulf* (Baltimore, MD: Nautical and Aviation Publishing Company of America, 1996), 86.

2. USCENTAF/JRCC Incident/Mission Log, DS box, Joint Personnel Recovery Agency (Fort Belvoir, VA, undated), (hereafter USCENTAF/JRCC Incident/Mission Log).

3. Stan Morse, ed., "Gulf Air War Debrief," (hereafter GAWD), *World Air Power Journal* (London, UK: Aerospace Publishing, 1991), 118; e-mails from Flt Lt Mary Hudson, RAF/AHB, 21, 23, and 27 March 2006; and http://www.raf.mod.uk/gulf/loss.html.

4. Charles Allen, *Thunder and Lightning: The RAF in the Gulf: Personal Experiences of War* (London, UK: Her Majesty's Stationery Office, 1991), 59.

5. Edward Marolda and Robert Schneller Jr., *Shield and Sword* (Washington, DC: Naval Historical Center, Department of the Navy, 1998), 236.

6. USCENTAF/JRCC Incident/Mission Log.

7. Center for Naval Analysis (CNA), "Desert Storm Reconstruction Report, Volume II: Strike Warfare" (Arlington, VA: October 1991), 5-58; and Morse, GAWD, 228.

8. USCENTAF/JRCC Incident/Mission Log.

9. Ibid.

10. Thomas A. Keaney and Eliot A. Cohen, *Gulf War Air Power Survey, Summary Report*, vol. 5, pt. 2 (hereafter *GWAPS*) (Washington, DC: US G.P.O., 1993), 184.

11. Murray with Thompson, *Air War in the Persian Gulf*, 183.

12. Ibid., 188.

13. Ibid., 198.

14. History, 1 Special Operations Wing (1 SOW), 1 January–31 December 1991 (History Office), 53; Chris Ryan, *The One That Got Away* (London, UK: Century Press, 1995), 201; and Capt William LeMenager, "A Gulf War Chronicle," (HQ SOCOM, unpublished manuscript, January 1998).

15. Marolda and Schneller Jr., *Shield and Sword*, 210.

16. Comments by Maj Gen Richard Comer on this book, 2 February 2003.

17. USCENTAF/JRCC Incident/Mission Log 15; History, 3d Marine Air Wing (3 MAW), and Marine Attack Squadron VMA-311 (Washington, DC: Marine Corps Historical Center, 1 January–28 February 1991).

18. History, VMA311.

19. History, 3 MAW, 119.

20. USCENTAF/JRCC Incident/Mission Log.

21. *GWAPS*, 184.

22. Michael R. Gordon and Gen Bernard E. Trainor, *The Generals' War: The Inside Story of the Conflict in the Gulf* (New York: Little, Brown and Co., 1994), 269.

23. Ibid., 271.

24. Murray with Thompson, *Air War in the Persian Gulf,* 252.
25. Ibid., 254; and Marolda and Schneller, *Shield and Sword,* 229.
26. USCENTAF/JRCC Incident/Mission Log.
27. Tom Clancy with Gen Chuck Horner, *Every Man a Tiger* (New York: G. P. Putnam's Sons, 1999), 450.
28. Morse, *GAWD,* 227.
29. Murray with Thompson, *Air War in the Persian Gulf,* 253.
30. Gordon and Trainor, *The Generals' War,* 285.
31. Ibid., 244.
32. Marolda and Schneller, *Shield and Sword,* 231.
33. Message, 021200Z FEB 91, CENTAF.
34. Sherman Baldwin, *Ironclaw: A Navy Pilot's Gulf War Experience* (New York: William Morrow, 1996), 194.
35. USCENTAF/JRCC Incident/Mission Log.
36. CNA, 5-48, 5-58; and Morse, "GAWD," 227.
37. Amy W. Yarsinske, *No One Left Behind: The Lt Cdr Michael Scott Speicher Story* (New York: Dutton, 2002), 99.
38. USCENTAF/JRCC Incident/Mission Log.
39. William L. Smallwood, *Warthogs: Flying the A-10 in the Gulf War* (Washington, DC: Brassey's Press, 1993), 141.
40. Lt Col Joe Hampton, interview by the author, 12 March 2000.
41. Smallwood, *Warthogs,* 142.
42. USCENTAF/JRCC Incident/Mission Log.
43. History, Marine Attack Helicopter Squadron 775 (HMLA-775); and History, 3 MAW, 150.
44. USCENTAF/JRCC Incident/Mission Log.
45. Message, JRCC A/R 031400Z FEB 91; and Clancy with Horner, *Every Man a Tiger,* 457.
46. Lt Col Mark A. Welsh, "Day of the Killer Scouts," *Air Force Magazine,* April 1993.
47. USCENTAF/JRCC Incident/Mission Log.
48. History, 3 MAW, 150.
49. USCENTAF/JRCC Incident/Mission Log.
50. CNA, 5-59; and Morse, "GAWD," 227.
51. Murray with Thompson, *Air War in the Persian Gulf,* 189.
52. Gordon and Trainor, *Generals' War,* 323.
53. Tim Ripley, "Desert Weasels," United States Air Force Yearbook 1992, 60.
54. USCENTAF/JRCC Incident/Mission Log.
55. Message, JRCC A/R 071400Z FEB 91.
56. Murray with Thompson, *Air War in the Persian Gulf,* 258.
57. BG Robert H. Scales, *Certain Victory: The US Army in the Gulf War* (Washington, DC: Office of the Chief of Staff, US Army, 1993), 186; and Gordon and Trainor, *Generals' War,* 246.
58. Tom Clancy with Gen Carl Stiner, *Shadow Warriors* (New York: G. P. Putnam's Sons, 2002), 439.
59. BG Robert H. Scales, *Certain Victory,* 186.
60. Col George Gray, interview by the author, 3 May 2001.
61. Capt Paul Harmon, interview by the author, 16 February 2000.
62. Comments by Maj Gen Rich Comer on this book, 2 February 2003.

63. USCENTAF/JRCC Incident/Mission Log.
64. History, Marine Attack Squadron, VMA 231; and History, 3 MAW, 133.
65. Gordon and Trainor, *Generals' War*, 307.
66. USCENTAF/JRCC Incident/Mission Log.
67. Naval Historical Center, Detachment Summary Report HCS-4, Helicopter Combat Support Special Squadron 4; and LCDR Neil Kinnear, interview by the author, 24 March 2001.
68. USCENTAF/JRCC Incident/Mission Log.
69. Message, JRCC A/R 141405Z FEB 91.
70. USCENTAF/JRCC Incident/Mission Log.
71. Clancy with Horner, *Every Man a Tiger*, 564.
72. Morse, GAWD, 227.
73. Detachment Summary Report, Helicopter Combat Support Special Squadron 4, (HCS-4), Desert Shield/Desert Storm (CNA, Arlington, VA: 9 December 1990–20 March 1991). Includes attached aircraft and personnel from HCS-5.
74. Yarsinske, *No One Left Behind*, 67.
75. Clancy with Horner, *Every Man a Tiger*, 564.
76. Murray with Thompson, *Air War in the Persian Gulf*, 179.

Chapter 6

Desert Storm Week Five: 14–20 February 1991

We knew they were around there somewhere, but we had been shot at a lot by missiles and AAA. I had flown 30 sorties, and I was feeling a little indestructible.

—1st Lt Robert Sweet

14 February

Desert Storm coalition commanders awoke to endless pictures on worldwide news outlets of Iraqi civilians killed when the Al Firdos bunker complex was bombed by F-117s the previous night. An immediate investigation was launched. Intelligence sources indicated that significant Iraqi leaders were sequestered in a command bunker there, but built above it was a bomb shelter that was full of civilians.

The horrific pictures brought immediate international condemnation upon the coalition effort. Responding to the negative publicity, General Schwarzkopf directed that no more targets could be hit in the Baghdad area without his direct approval.[1] Otherwise, the aerial onslaught against the Iraqi forces continued unabated.

I072 – Belfast 41/Tornado, RAF

A combined flight of eight Tornadoes and four Buccaneers carrying laser-guided bombs again attacked the Iraqi airfield at Al Taqaddum.[2] As they approached the target, several SAM sites began to track them. One locked on to the lead aircraft and fired several missiles. A British voice called, "Two missile launches!" just as two SA-2 radar-guided missiles slammed into the lead aircraft. The crew ejected.

The loss was reported to the JRCC, who tasked it to SOCCENT. The personnel in the SOCCENT RCC began to prepare a mission and also queried Proven Force about the possibility of

DESERT STORM WEEK FIVE

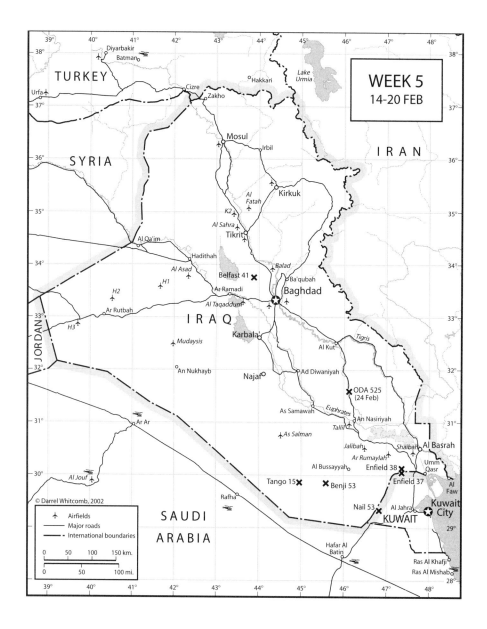

running a mission from the north. Since neither voice nor beeper contact was made with either crewmember, both SOCCENT and Proven Force requested that all flights into the area monitor the rescue frequencies for any signals. This was difficult because of a radio transmitter in Jordan that was jamming the rescue frequencies. Contact was never made with either man. Subsequently, intelligence sources determined that the pilot, Flt Lt Rupert Clark, had ejected and was immediately captured. The navigator, Flt Lt Stephen Hicks, was killed in the engagement.[3]

15 February

As the air campaign continued, Saddam again tried to strike back by reaching out with his Scud missiles, launching again into Saudi Arabia. One missile streaked southward and was not intercepted by Patriot batteries. It hit the water just 20 yards from a pier in the Al Jubayl harbor but did not explode. Several large ships were in port, including the US Navy amphibious ship *Tarawa* and a Polish hospital ship. Five thousand tons of artillery ammunition were stacked on the dock.[4]

In the KTO, Saddam's military forces were being steadily weakened but were still dangerous. This was brutally demonstrated to the tactical aircrews in a dramatic way.

At the end of January, General Glosson approved an operational change allowing the A-10s to operate as low as 4,000 feet unless the enemy threat clearly dictated otherwise. Up to that point, only three aircraft had been damaged by enemy fire. Flying lower, they were much more accurate with their weapons, especially the 30 mm gun that was their signature weapon.

Since that time 40 more aircraft had been damaged and one, call sign Uzi 11, shot down. With the focus now on Iraqi ground units, the A-10s were being sent deeper into Kuwait and Iraq and attacking dangerous and well-defended units like the Medina and Tawakina Divisions. But the sturdy A-10s had been built to take damage from enemy fire, and all except one had been able to fly back to their home stations.

IO73/MO16 – Enfield 37 and 38/A-10s, USAF

Numerous flights of two A-10s each launched to hit elements of the Iraqi 17th Armored Division in one of the kill boxes in Kuwait. Enfield 37 and 38 coordinated with other flights to hit an armored battalion.[5] They were especially alert because an earlier flight reported several missiles fired at them. While orbiting to spot targets, the wingman, 1st Lt Bob Sweet, observed ground fire. His flight lead, Capt Steve Phylis, rolled in and strafed it. Sweet began to follow when his aircraft was hit by an undetected heat-seeking missile, which struck his right engine. The blast destroyed the engine and severely damaged the right wing. Sweet immediately turned south and tried to reach the border. The blast had ripped open his hydraulic lines. As his hydraulic fluid dissipated, his flight controls froze, and the aircraft rolled out of control. The aircraft nosed downward, and he tried to switch the flight controls to the standby manual mode so that he could control the aircraft without hydraulic power. That process takes several seconds, and as the aircraft increased its rate of dive, Sweet realized that he did not have the time. He made a quick radio call that he was getting out and, unsure if his flight lead had heard him, pulled the ejection handle. He had been hit at about 12,000 feet and was well below 6,000 feet when he ejected.[6]

Descending, he saw enemy troops running toward him. He landed about 30 yards from an Iraqi headquarters bunker and some tanks. About 60 soldiers surrounded him and took him prisoner. He was never able to make an emergency call.[7]

The flight lead, Capt Steve Phylis, immediately took over as the on-scene commander for the rescue effort. He contacted another flight of two A-10s, Pachmayr 03 and 04, who were working just to the north and had them divert. As they approached, Phylis began giving them a verbal description of the area. They did not have visual contact with him. Suddenly he called, "This is Enfield 37, I am bagged at this time also." This was the code word for shot down. The flight lead Pachmayr 03 called for confirmation, but there was no response. He took over as the on-scene commander for the rescue of now two downed A-10s.

Arriving over the suspected positions, the pilots could not see any A-10 wreckage. As they searched, they were also shot at by heat-seeking missiles. Pachmayr 03 immediately contacted AWACS and tried to coordinate a search. AWACS notified the JRCC, and they tasked the recovery to SOCCENT. SOCCENT immediately began planning a mission, but with no voice contact or location, and considering the enemy forces in the area, the mission was not committed. Later, intelligence sources determined that Sweet had been captured and Phylis had been killed.[8]

These were not the only enemy successes that day. They had fired a total of eight observed heat-seeking missiles. Another struck the aircraft of the A-10 wing commander himself. His aircraft was severely damaged, but he was able to fly home. Crossing the border he noticed a flight of F-16s hitting targets just north of the line. Returning to base, he called Glosson and challenged the efficacy of sending the A-10s deep and the F-16s shallow when the F-16s were much more capable of surviving in the high-threat areas. Glosson took the matter to Horner, who decided to restrict the A-10s to targets along the border. Additionally, he directed that they had to stay above 10,000 feet and not use the 30 mm gun. This dramatically decreased their effectiveness, but it also reduced the loss rate. Besides, the F-111s were now flying in the area steadily, and their tank plinking was dramatically reducing the number of usable Iraqi tanks.[9]

16 February

I077 – UH-60, US Army

The aircraft was on a routine resupply mission and crashed 60 miles southeast of Rafha.[10] The Army RCC responded and sent a recovery aircraft. The crew was recovered uninjured. This was a noncombat loss.[11]

17 February

Since early February, targets in the Baghdad area had only been attacked by cruise missiles. When a missile was captured on CNN video being shot down, General Powell ordered the cessation of all daytime cruise missile launches. Combined with the restrictions on F-117 missions, this meant that targets in the Baghdad area were no longer being hit.[12]

Coalition air units had also targeted Iraqi bridges, but there was never a sustained interdiction campaign that came close to matching the effort being put forth to destroy the enemy units in Kuwait. The focus of the interdiction was to isolate the battlefield.

At SOCCENT, Colonel Gray still had operational control of the 3d/160th Battalion, the Army special operations helicopter unit from Fort Campbell, Kentucky. Highly trained, they were used to working with special operations ground teams and were providing steady support to the JSOTF operations in western Iraq. Due to the priority of combat recovery, they were also ordered to have aircraft on alert at Rafha and KKMC for immediate launch in the central CSAR sector. By this late date, they still had not gotten a CSAR mission and were tired of sitting alert. They were chafing at the inaction and beginning to wonder about the sense of it all.

I078/M017 – Benji 53/F-16, USAF

A flight of two F-16s was dispatched on a reconnaissance mission along the Euphrates River.[13] As they turned for home, the engine of the lead aircraft flown by Capt Scott Thomas began to fail. He notified AWACS and asked for an immediate vector to friendly territory. The AWACS controller acknowledged his call, stated that he had radar contact with the flight, and gave Thomas a heading.

Thomas also asked that JRCC be informed of his dilemma. The AWACS controller responded that rescue helicopters were already airborne. As they headed toward Saudi airspace, the wingman, 1st Lt Eric Dodson (Benji 54), asked for confirmation that the rescue forces were airborne. The AWACS controller reaffirmed that they were but said that he would double-check. As he was doing so, the engine of Benji 53 seized, and he was forced to eject.

Benji 54 stayed over his flight lead and became on-scene commander. He saw a good parachute on the ground, then Thomas contacted him on his PRC-90. Dodson had him switch to the secondary frequency. Making the change, Thomas stated that he was down and in good condition.

Dodson reported all of this to the AWACS controller. He could stay over Benji 53 for only another 15 minutes and asked for a relief flight. AWACS said they had some A-10s inbound. Dodson asked where they were and was told that they were 200 miles south—too far out to relieve him before he had to leave. He asked that a closer flight be diverted to cover the survivor. A flight of F-15Es was brought in, and Dodson handed the situation off to them and headed for the F-16 FOL (forward operating location) at KKMC.

Landing at KKMC, Dodson called the Army SOF guys across the field to see how the rescue effort was going for his flight lead. They told him that contrary to what AWACS had reported, no rescue forces had been sent yet because they had not been told that positive contact had been established with the pilot.[14]

Armed with this new information and the position passed earlier by Dodson, they checked with SOCCENT and were cleared to take the mission. The unit there at KKMC was the 3d/160th under the command of LTC Dell Daily. He and several crews had just returned from a combat recovery practice mission. He directed one of the young pilots, CW3 Tom Montgomery, to start planning the mission.

Montgomery jumped to it. Taking the pilot's reported position, he and his copilot, CW3 Joel Locks, went into Intelligence and began planning a route. Initially two aircraft were available, so they planned to use both in formation. As they tried to start the engines, the second aircraft broke, so Montgomery pressed on alone. It was dark, and he felt that, despite the enemy threat, he would be safe enough.

As Montgomery approached the border, the second aircraft was able to launch. The pilot, CW3 Todd Thelin, with copilot, CW3 John Aberg, called Montgomery and said that they would catch him at maximum speed. They joined just as they entered Iraq. Both crews were flying using NVGs. Flying at 15–20 feet, their planned route took them around enemy forces in the area.

Montgomery contacted AWACS, who tracked them as they made their way into the area. He decided that the safest way to approach Thomas was from the north, so he flew by and made a right fishhook to fly directly to him.

Hearing the helicopter, Thomas took out his radio and tried to call them. They were still at about 15 feet altitude and did not hear him. Overflying the predicted position, Montgomery called AWACS for a vector. AWACS gave him a precise heading and distance to the survivor.

Thomas heard AWACS turn the helicopter. He took out his infrared strobe light and turned it on. It began flashing perfectly. Montgomery said,

I see a strobe-light. It was pitch dark. I could hardly believe it. . . . So I called to my wingman, "Hey, there is a strobe-light at 12 o'clock." I looked at my GPS—6.7 miles. It's amazing. So we did an over-flight, saw people and trucks on the ground about 500 meters away. My wingman said that he had people and trucks on his side about a click [kilometer] away. They [the enemy] lit us up with radar, the RWR [radar-warning receiver] was lit up. . . . We did a right turn and landed down wind, no big deal. Thomas was standing there like he was waiting for a bus. The only thing that saved his butt was that strobe light. That is what we saw. I landed the aircraft three feet away.[15]

As the helicopter touched down, Thomas fell into the fetal position. The crew chief and two Army Special Forces troops jumped off and grabbed him. They put him on board, reloaded themselves, and Montgomery took off. As they cleared the area, they notified AWACS, who then cleared the fighters capping the operation to hit the enemy troops. They covered the area with cluster bombs.

As Montgomery flew outbound with his survivor on board, the Iraqi radars continued to track them, at one point even firing a missile, but it did not guide on the aircraft. He and his wingman stayed at 15 feet until they entered Saudi airspace, then they climbed and headed for KKMC.[16]

In the JRCC, Colonel Hampton was pleased with the rescue. In analyzing the mission, he remembered that, "We got some good intelligence. We determined that there was not a heck of a lot around there, not a high military concentration area, so we were able to get assets across the border from where they launched. They were able to get up there quickly and pick the guy up before he was compromised."[17]

It was a quick and successful rescue by the Army SOF. After that, the men of the 3d/160th had a different attitude about sitting CSAR alert—no more complaining. Col Ben Orrell said of their metamorphosis, "It was interesting because the night before the [Benji 53] mission occurred, they came to us and said that they were sick and tired of this, that they were not being used for anything and were just sitting there wasting their time and thought they should be pulled out. I said, 'Stay on until the end of the week and I'll see if I can get you out of it.' They got that mission, got to go up there and came back and said, 'You know, we think we like this!'"[18]

19 February

A large strike package from Proven Force was launched out of Turkey and struck targets in the Baghdad area. Flying as a composite force, they preceded the strike with integrated Wild Weasel attacks against several Iraqi SAM sites. As a result, all targets were struck, and no aircraft were shot down.

Coalition aircraft were also active farther south. So were the enemy gunners.

I079 – Nail 53 OA-10, USAF

Flying a forward air controller mission, this aircraft entered one of the kill boxes in Iraq.[19] Lt Col Jeff Fox, the squadron operations officer, relieved his predecessor on station and began scanning for targets. His assigned priority was artillery, tanks, and other vehicles. Entering the area at 17,000 feet, he gradually descended as he searched for targets. Realizing that he was down to 8,000 feet, he added power to climb back to altitude when he felt the aircraft shudder from a solid thump underneath. He was headed south. For a moment, all seemed normal. Then warning lights in the cockpit began to illuminate. The flight controls became unresponsive, and the nose pitched up slightly. Fox immediately noticed that his hydraulic pressure had fallen to zero. He switched the flight controls to manual reversion and made some emergency radio calls, stating his position and intentions. Several aircraft answered and began proceeding to his location.[20]

Manual reversion had no effect, and the aircraft rolled over and began to accelerate toward the ground. With no control of the aircraft, Fox ejected. His chute opened and he heard his aircraft explode when it hit the ground. The 30 mm rounds in the gun began to explode. His right heel, elbow, and knee had been injured in the ejection. He landed near the burning wreckage and immediately moved away from the site.

AWACS monitored his emergency calls and reported his downing to the JRCC. Since he had been working in his assigned kill box, they had an approximate location. He was across the border from the VII Corps, so the JRCC contacted the Army RCC to see if they could help. The Army would not launch a helicopter because Fox was too deep in enemy territory. The JRCC did get two A-10 Sandy aircraft scrambled.

On the ground, Fox took out his PRC-112 survival radio and began calling for help. Several flights answered and headed for his location. Another forward air controller spotted the smoke of his burning aircraft and flew to him. The pilot asked for Fox's position from the smoke. "One hundred meters south," Fox replied, as the OA-10 began to orbit overhead.

Unfortunately, Fox had been observed by enemy forces as he descended. An Iraqi soldier carrying a rifle rapidly approached shouting in Arabic and firing his rifle at Fox. The wounded pilot tried to run away, but the soldier closed on him. The fighters began arriving overhead and called Fox for direction. Fox eyed the Iraqi soldier who was now just about upon him. Realizing that the race was lost, Fox keyed his radio and said, "Okay, it's too late guys, he's here." He was taken prisoner.[21] With the loss of voice contact, and his location well within enemy units, any thought of a rescue was abandoned.

***

Watching the imminent approach of the ground battle phase of the war, Brig Gen Buster Glosson decided to communicate again with all of his crews. Given the level of activity by all, visiting the bases again was impractical. Instead, he sent a message to all units. It said:

> All of you heard me say earlier that not one thing in Iraq is worth dying for, and that was true. Sometime in the next week, there is going to be a lot worth dying for in Iraq. We call them American soldiers and marines. When I said that I wanted a minimum loss of life, we cannot draw distinctions between Americans who die. If a marine dies or a soldier dies, it's all the same to me as one of you dying. For that reason, there will be no restrictions placed on you by anyone. The individual fighter pilot or flight lead will decide what is necessary.[22]

Going into the ground war, it was "no holds barred." The aircrews could lay it all on the line.

20 February

An F-16, Barretta 11, from Proven Force could not refuel from a tanker and tried to divert to Diyarbakur, Turkey. He flamed out on short final and ejected. Local Turkish forces recovered the pilot. This was the only aircraft lost in the northern operation.[23]

In the Persian Gulf, a small but historically significant force entered the fight. AV-8s from VMA-331 aboard the amphibious ship USS *Nassau* (LHA-4) launched to strike Iraqi targets. This was the first time ever that vertical-assault aircraft had taken off from a helicopter landing ship and struck enemy targets. Before the end of hostilities, the squadron would fly 243 interdiction and close air support sorties.[24]

As G-day (beginning of the ground attack) approached, the pace was frantic for the ground units. Their helicopters were constantly in the air.

I080 – Lady Ace 03/CH-46, USMC

This aircraft from Marine Medium Helicopter Squadron 165 was returning from a night priority mission when it encountered a fog bank.[25] While attempting to land in the fog and blowing sand, the aircraft touched down in a sideward drift and crashed. Fuel tanks ruptured and the aircraft was destroyed. The crew and passengers escaped uninjured. This was subsequently determined to be a noncombat loss.[26] More importantly, though, it illustrated the difficulty of helicopter operations in the region. The combination of fog and blowing sand was a dangerous condition that could cause terrible disorientation for the aircrews during the most critical phase of flight.

I081 – Tango 15/OH-58, US Army

The aircraft from the 24th Infantry Division was part of a package of Scouts and Apaches that was performing a night reconnaissance mission deep in Iraq.[27] It went down in bad weather 55 miles southeast of Salman Air Base, Iraq. An initial search was unsuccessful, and the Army RCC requested help from the JRCC. Aircraft were diverted into the area and started an electronic search, but no beepers were heard, and no voice contact was made. SOCCENT was contacted and began planning a mission pending voice contact. The Army RCC set up a restricted operating area around the last-known position for the crew. The next morning, they dispatched a helicopter task force on a search mission. They found both crewmembers dead in the aircraft wreckage.[28]

Special Forces Team ODA 525

This was not a CSAR action in the strictest sense—the Army calls such missions "emergency exfiltrations"—but the exploits of Army Special Forces team ODA 525 approximate one in almost every way. This team from the 1st Battalion of the 5th Special Forces Group was inserted into Iraq on a strategic reconnaissance mission in the early hours of 24 February. The eight-man team was carried in by two MH-60s from the 3d/160th, the same guys who had rescued Capt Scott Thomas (Benji 53) eight days prior.

Each team member carried 150 pounds of equipment as the helicopters flew them deep into Iraq. Flying for almost two hours, they were dropped off about 30 miles north of Nasiriyah along Highway 8. Their assigned task was to monitor road traffic in direct support of the XVIII Airborne Corps that would soon be sweeping into Iraq.

Team chief CW2 Richard Balwanz's orders from the Corps commander were direct. "He wanted to know what the enemy was doing in that area," Balwanz recalled. "He wanted to know what kind of equipment was coming down that highway, or leaving." He wanted HUMINT—human intelligence—guys on the ground to observe for him. The men of ODA 525 were his guys.[29]

But the infiltration did not go as expected. As the two helicopters entered Iraqi airspace, they were contacted and told to abort the mission and return to base. Turning around, they were redirected to land at Rafha and refuel, pending a change of mission. As the helicopters fueled, the team was again ordered to execute. The two helicopters lifted off and headed for the assigned position.

Unfortunately, the delay caused a navigation problem. The helicopters were using GPS to navigate. As they approached the hide site, their GPS lost contact with its satellites and stopped updating. Even with 16 satellites now in orbit and providing specific passage over the Gulf region, there were still occasional gaps in coverage that caused tactical problems for the warriors below who had become used to the ease, convenience, and precision of GPS navigation. As backup, they had to navigate by compass and map. In doing so, the pilot of the lead aircraft got too low and hit a sand dune. This knocked off one of the landing wheels. Unfazed, the crew proceeded. As they approached the drop-off point, the pilots stopped at two other locations as insertion decoys just in case any enemy troops were watching.

Landing at the estimated site, the team hopped off into the clear night. Balwanz did a quick head count and waited to ensure that they had not been observed. He did not want to let the helicopters leave if they had been compromised. When all was in order, he let them go. The team was alone.

Balwanz turned on his GPS receiver. The satellites were back in position, and he was able to get a precise fix on their location. They were 1,500 meters north of where they were supposed to be. The team started moving. Arriving at their assigned position, Balwanz decided to split his team into two sections and locate them so that each had an excellent view of the highway just 300 meters away. The men started to dig their hide sites. The ground was hard, and they finished just as the sun was rising.

It was not too soon, because 30 minutes later, civilians began to come into the fields. Some were there to work the crops. Some were gathering wood. Some were children playing. They were accompanied by grazing animals that moved toward the team.

The men crouched in their holes, but to their horror, two small children wandered up to Balwanz's site. Curiously peering under the camouflage net, one of them said "enemy" in Arabic, then they screamed and ran away. The men immediately grabbed their weapons, but Balwanz did not tell anybody to shoot. "We are not going to shoot any unarmed civilians, especially children," he said. Instead, he grabbed the radio to call the other element. They had been compromised and they needed to leave.

As both groups rendezvoused in a nearby canal, Balwanz called Corps headquarters and asked for an exfiltration. He moved the team about 200 meters away, but still in a position where they could see the road, and waited. Balwanz could see other civilians, but there was nothing unusual in their activities.

Now convinced that the children had not raised any alarm, he decided to continue the mission and so notified headquarters.

For the next four hours, they monitored and reported the traffic. It was steady heading north but sparse heading south. Then he spotted some more children heading his way. This time, they were accompanied by an adult. As they approached, Balwanz stood up and greeted him in Arabic. The man mumbled something that they did not understand and, with the children, ran away. They watched him head into town.

Alarmed, Balwanz began to move his team away from the village. He spotted several unarmed young men in military uniforms moving toward him. He yelled at them in Arabic to stop. They were joined by a group of older men with antique rifles. Not far behind them were several trucks carrying an estimated 150 soldiers armed with AK-47s.

Balwanz called Corps headquarters again, gave them the teams updated GPS coordinates, and asked for an expedited emergency exfiltration. He also told them that he was about to be attacked by a company-size force. They replied that the helicopters were on the way.

As the soldiers moved to flank them, Balwanz had his team gather all their gear except for weapons and a few radios and rig it for destruction. They set the explosive and moved away from it. As the charge went off, the Iraqis started to shoot at them. Balwanz did not want to fight, but he had no choice. He cleared his men to engage the force. They started dropping the Iraqi soldiers and killed 40 in ten minutes.

The team could see that other soldiers had run over to the road and were flagging down military vehicles. More enemy were joining the fight. The team needed some heavier firepower and fast. When the Corps headquarters launched the helicopters, they contacted AWACS and had them send an airborne forward air controller to the team's location. The forward air controller, call sign Pointer, orbited over their position and tried to talk with them. The team's primary radio had failed. One of the soldiers had a PRC-90 survival radio. He turned it on and was able to make contact with Pointer. When the forward air controller had a feel for the situation, he started calling for air support. AWACS started diverting strike flights to him.

DESERT STORM WEEK FIVE

Photo courtesy of Bill Bryan

Capt Bill Andrews, *far right,* **led the successful rescue of ODA 525 from deep inside Iraq. He was later the subject of a CSAR himself as Mutt 41. He is shown here with other members of Mutt flight,** *left to right,* **1st Lt Pete McCaffrey, Capt Evan Thomas, and 1st Lt Joey Booher.**

The first flight was four F-16s. They were carrying CBU-87 cluster bombs. Such weapons are terribly destructive and totally indiscriminate. Before anything could be dropped, the forward air controller needed to make sure that the strike pilots could see the team. That is difficult on the best of days, but using some of the natural landmarks, the ground team was able to "talk" the forward air controller and strike pilots onto their location so that they could drop their ordnance without harming the team. The F-16s dropped their CBUs on the vehicles and troops. A following convoy of troops saw the destruction and wisely declined to join the battle.

The team redirected the F-16s to more enemy forces. The second flight dropped more CBUs with similar results. The forward air controller then diverted them to still another gathering group of enemy. That had the same result. The immediate threat was eliminated.

Balwanz spotted three soldiers who had survived the air strike. He personally shot them. He then had the F-16s hit some more vehicles and soldiers on the road before they had to leave.

By now the forward air controller was low on fuel and had to depart. He handed off the fight to four F-16s from the 10th TFS. They were led by Capt Bill Andrews. Before letting Pointer leave, Andrews had the team flash a mirror at them so he had a positive fix on their location. Then, he and his flight continued to ring the team with explosive steel.[30]

Andrews and his flight stayed over the men as the day grew late. Repeatedly, the team called for the CBU. The pilots dropped it as close as they thought safe. The team demanded it even closer. The pilots complied, until the bomblets were exploding less than 100 meters from Balwanz and his men. It broke the enemy attack.[31]

With the sun setting, Balwanz moved his team west about 300 meters to a better landing zone for the helicopters. It also put some distance between them and the enemy force. There, another F-16 overhead told them that the helicopters were just 10 minutes out.

When the MH-60 helicopter crew contacted them on their PRC-90, Balwanz transmitted a homing signal that the rescue crew was able to follow for the last mile. They came in and landed right next to the team. As the F-16s orbited protectively overhead, Balwanz and his men scrambled on board and headed for home.[32]

Landing back at their base, Andrews and his flight mates jumped in a jeep and drove over to the intelligence shop. They asked the officer there if he could check on the status of ODA 525. He made a few calls and was able to inform them that all eight men had made it out safe and unharmed. It was a successful rescue.[33]

There were other teams, also. While not counted as CSARs, their rescues were similar. It was the same helicopter crews, using the same techniques. Throughout the war, teams of US Army, Navy, Marine, and coalition forces operated throughout Iraq and Kuwait. Most of their stories are untold. Most are still classified. But it was the flyers of SOCCENT, the Army, Air Force, and Navy who took them in and brought them out. Someday, their stories will be told, as well.

Again, the ODA 525 recovery showed the absolute efficacy of GPS. It was GPS that enabled the relatively rapid dispatch of the helicopters. Knowing exactly where they needed to go, they were able to actually preplan the mission, only pausing for last-minute updates before launching for the recovery. A postwar report explained how basically simple this was using GPS. It said that, "with start and end points available to the user, navigation could be accomplished without the need for charts or observation of any object or reference point other than the receiver display."[34] The key was in having a known "end point."

The absolute value of GPS was also highlighted in another small incident that, in the larger swirl of the war, was probably not noted by anybody at the command level. It occurred on a Navy SEAL mission that took place off the coast of Kuwait at roughly the same time as the Balwanz mission. The small team had completed its tasks and was awaiting extraction. As the recovery vehicle approached, the team inadvertently dropped a valuable piece of equipment into the water. They could not delay their pickup to retrieve it. Instead, they marked the position with their GPS receiver. The next night, using their GPS for navigation, they slipped back in and recovered it. The report went on to state that, "The [SEAL] team's vulnerability to exposure (surveillance or fire) was directly reduced by GPS navigation accuracy."[35]

It was a simple act, a small footnote to a much larger war. But these two events showed the potential that GPS offered for safely recovering isolated personnel behind enemy lines.

Notes

1. Murray Williamson with Wayne Thompson, *Air War in the Persian Gulf* (Baltimore, MD: Nautical and Aviation Publishing Company of America, 1996), 204.

2. USCENTAF/JRCC Incident/Mission Log, DS box, Joint Personnel Recovery Agency, Fort Belvoir, VA, n.d. [hereafter USCENTAF/JRCC Incident/Mission Log].

3. Stan Morse, ed., "Gulf Air War Debrief" (hereafter GAWD), *World Air Power Journal* (London, UK: Aerospace Publishing, 1991), 227; and Charles Allen, *Thunder and Lightning: The RAF in the Gulf: Personal Experiences of War* (London, UK: Her Majesty's Stationery Office, 1991), 117.

4. Edward Marolda and Robert Schneller Jr., S*hield and Sword* (Washington, DC: Naval Historical Center, Department of the Navy, 1998), 197.

5. USCENTAF/JRCC Incident/Mission Log.

6. Robert J. Sweet, oral history interview (Colorado Springs, CO: Air Force Academy, 28 June 1991).

7. Ibid.

8. Tape of mission of Pachmayr 01 flight on 15 February 1991; and William L. Smallwood, *Warthogs: Flying the A-10 in the Gulf War* (Washington, DC: Brassey's Press, 1993), 178–80.

9. Murray with Thompson, *Air War in the Persian Gulf*, 259.

10. USCENTAF/JRCC Incident/Mission Log.

11. Message, JRCC A/R 171430Z FEB 91.

12. Michael R. Gordon and Gen Bernard E. Trainor, *The Generals' War: The Inside Story of the Conflict in the Gulf* (New York: Little, Brown and Co., 1994), 327.

13. USCENTAF/JRCC Incident/Mission Log.

14. William L. Smallwood, *Strike Eagle: Flying the F-15E in the Gulf War* (Washington, DC: Brassey's Press, 1994), 126.

15. CW3 Tom Montgomery, interview by the author, 22 July 2000. The crew consisted of: CW3 Tom Montgomery; CW3 Joel Locks; SFC Paul Laduca; SGT Kurt Hixenbaugh; PFC William Mudd; SFC Edmund Wilson; and SSG Douglas Patterson.

16. Lt Gen E. M. Flanagan, "Hostile Territory Was Their AO in Desert Storm," *Army*, September 1991, 12; and Montgomery, interview.

17. Lt Col Joe Hampton, interview by the author, 12 March 2000.

18. Col Ben Orrell, interview by the author, 17 January 2002.

19. USCENTAF/JRCC Incident/Mission Log.

20. Smallwood, *Warthogs*, 184.

21. Ibid., 185.

22. Al Santoli, *Leading the Way: How Vietnam Veterans Rebuilt the U.S. Military: An Oral History* (New York: Ballantine Books, 1993), 205.

23. Desert Storm Narrative log, JPRA/Desert Storm file, box 15c1.

24. Marolda and Schneller, *Shield and Sword*, 241.

25. USCENTAF/JRCC Incident/Mission Log.

26. History, 3d Marine Air Wing (3 MAW), Marine Attack Squadron VMA-311, (Washington, DC: Marine Corps Historical Center, 1 January–28 February 1991), 150.

27. USCENTAF/JRCC Incident/Mission Log.

28. Message, RCC A/R 211430Z FEB 91, ARCENT G-3 AVIATION, After Action Summary, 5 April 1991; and James W. Bradin, *From Hot Air to Hellfire: The History of Army Attack Aviation* (Novato, CA: Presidio Press, 1994), 185.

29. Dale B. Cooper, "Bulldog Balwanz and his Eight Man Army," *Soldier of Fortune*, May 1992, 32.

30. Edward Herlick, "Daring Rescue Deep in Iraq," *Military History*, December 1994, 62.

31. Ibid.

32. Cooper, "Bulldog Balwanz," 32.

33. Herlick, "Daring Rescue," 62.

34. CDR Patrick Sharrett et al., "GPS Performance: An Initial Assessment," *Navigation Journal*, vol. 39, no. 1 (Spring 1992), 402.

35. Ibid.

Chapter 7

Desert Sabre Week Six: 21–28 February 1991

War is still, when you get to the bottom line, killing, maiming, and wanton destruction.

—Lt Gen Chuck Horner

21 February

Delta Force commandos in the JSOTF, working in western Iraq, stumbled into a major firefight, took some casualties, and requested a medical evacuation. An MH-60 from the 3d/160th flew in to get them. When it returned to ArAr, the weather was terrible, with fog severely limiting visibility. The pilots were unable to land in the fog and crashed, killing all aboard.[1] It was not listed as a combat loss.

For the pending ground campaign, almost all air effort was refocused on the Kuwaiti theater. The F-111Fs were directed to hit Iraqi artillery. F-117s were refragged to hit key command and control centers for specific units.[2]

In preparation for the ground war, aviation units from the XVIII Airborne Corps began operations into southern Iraq. One helicopter operation attacked a fortified bunker and took 406 prisoners to include a battalion commander.[3]

Elements of the 1st Marine Division began infiltrating into Kuwait in preparation for the ground war, now scheduled to begin on the 24th. The marines found the enemy resistance surprisingly light. They were able to penetrate the first lines of enemy wire and mines. In the process, they began to collect significant numbers of prisoners.[4]

23 February

In the early morning hours the 2d Marine Division began cross-border operations in preparation for the main attack. In one sharp engagement, they destroyed 20 Iraqi tanks and 25

DESERT SABRE, WEEK SIX

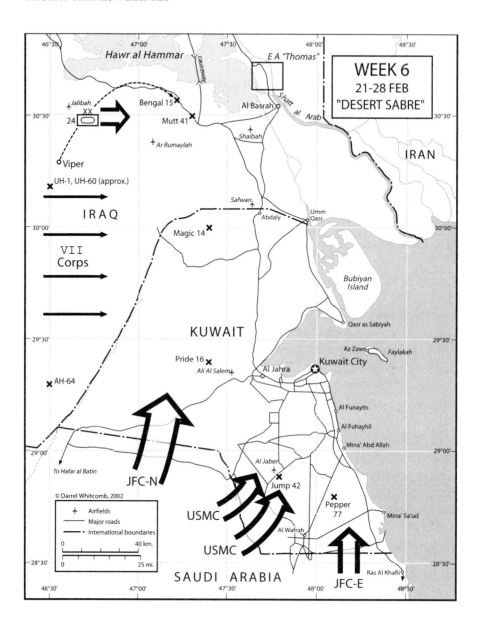

other vehicles and captured more than 300 prisoners.[5] Air support from Marine Cobra helicopters, AV-8s, A-6s, and F-18s and from other services was always overhead, but the Iraqi ground forces were well protected with SAMs and AAA.

210

I083/M018 – Pride 16/AV-8 Harrier, USMC

Capt James "LZ" Wilbourn, VMA-542, was reported lost on a night sortie five miles north of the Ali Al Salem Air Base.[6] Flying with Maj Dan Peters, he had already flown one mission over Kuwait. Returning with another load of bombs, they rendezvoused with a Marine A-6 that had a lucrative target.

Due to the darkness, the Harrier pilots made individual passes. Peters watched Wilbourn roll in on the target. He lost sight of the Harrier in the darkness but did observe the explosion of Wilbourn's bombs. This was followed almost immediately by another larger blast. Peters thought it was a secondary explosion, but when Wilbourn did not answer his radio calls, Peters realized that he was down. He initiated CSAR procedures, and throughout the night, crews in F-15Es, F-111s, and AV-8s searched for him.

Peters did not see what shot down Wilbourn. It was entirely possible that he became fixated on his target and flew into the ground.[7] The young captain was the first US marine actually confirmed as killed in the war.[8]

24 February

For the last several weeks, under the protective umbrella of US and coalition airpower, the ground forces had moved a massive two-corps force west to be in position to envelop the mass of Iraqi troops in Kuwait and strike a decisive blow against the Republican Guard units. The GPS system was the great enabler. Never before had such a force been able to accurately move across the desert in this manner. Indeed, Saddam and his commanders were surprised by the ability of the coalition forces to move as accurately and quickly as they did.[9] They thought that the desert was their best defensive feature because, heretofore, precise cross-country travel in the desert could only be accomplished using the few roads available. The Iraqi forces had them all covered with strong, entrenched forces. General Schwarzkopf would use his navigational advantage to go around them.

There was some risk involved in putting such reliance on the GPS system. Making it available to the maximum number of receivers meant some constraints and safeguards built into the system had to be relaxed. This made it available to anybody who procured commercially available receivers and locked onto the satellites. This most certainly included the Iraqis, but there is no evidence that they ever did so or even realized the value of the GPS system.[10]

This huge coalition maneuver to the west would be supported by a combined Arab and US Marine/Army attack directly into Kuwait as well as an amphibious demonstration off the coast designed to pin Iraqi forces on the beaches. CENTAF was prepared to support all the elements of the attack. At the afternoon staff meeting, General Horner told them, "There are people's lives depending on our ability to help them, if help is required. So I want a push on. I want people feeling compulsion to hit the target. I do not want fratricide . . . but up over the battlefield, it's time to go to work because other people's lives depend on ours. It's no longer a case of air just risking their own lives; other lives have to be considered."[11]

He took all restrictions off of his aircrews; they could fly as low as they felt necessary to accomplish the mission. There were two other dictating factors. First, a messy weather front moved in bringing with it low clouds, reduced visibility, and even icing at higher altitudes. This forced the pilots to go in below the clouds so that they could accurately see their targets for precision attacks. Second, the Iraqis started igniting the oil wellheads, creating huge petroleum fires with billowing clouds of thick smoke. This had an especially detrimental effect on the sensor and aiming systems of the aircraft.[12]

This change in tactics pushed the aircrews down into the guns and the heat-seeking missiles. One pilot described the danger as he returned from a target near Basra:

> The AAA was heavier than I had ever seen it. What we didn't know was that two Republican Guard divisions had moved onto the road along our route of travel. I will never, ever, forget what that looked like. Just a wall of AAA. Down low there was an illusion of going down a tunnel because the AAA just kind of parted in front of us and passed over the top of the aircraft. It was so thick I just squeezed down into my seat and waited to get hit. What else could I do? I couldn't turn around. I couldn't go left, couldn't go right. My whole philosophy was, "I'm going to get through this stuff as fast as I can."[13]

At 0100L the ground offensive, code-named Desert Sabre, started. On the far west, the French 6th Light Armored Division moved north to screen the entire left flank of the coalition force. To their right, the US Army 101st Airborne Division started a massive entire brigade air assault to a position just short of the Euphrates River. This mass of helicopters was covered by swarms of A-10s and F-16s, all in contact with the air liaison officers (ALO) with those units. They were the lead elements of the XVIII Airborne Corps.

Also assigned to the XVIII Corps were the 24th Mechanized Infantry Division and the 3d Armored Cavalry Regiment. Their tanks and armored personnel carriers were the last to step out behind the French units and American helicopter-borne forces.

To the right of them was the VII Corps with several heavy mechanized and armored divisions and a British armored division. Their specific target was the Republican Guard. Their mission was simple in statement—destroy the crème of Iraq's ground forces. Their attack would begin on the 25th.

To their right were the Arab forces of Joint Forces Command North (JFC-N). They would enter Kuwait with the two divisions of US Marine forces. One, the 2d Marine Division, had an Army armored brigade attached to it. Called the "Tiger" Brigade, it was a powerful addition to the Marine force. Along the coast was another smaller Arab force, Joint Forces Command East (JFC-E). Above them all swarmed masses of A-10s, F-16s, AV-8s, Tornadoes, Jaguars, F-18s, A-6s, F-15s, and countless helicopters ready to pounce on any intransigent enemy.

Under constant attack from the air, and now the massed artillery of the advancing coalition forces, Iraqi gunners were unable to shoot down any coalition aircraft. The biggest threat to the flyers was the possibility of mid-air collisions in the lousy weather.

The A-10s had been designed for this. They worked well below the clouds and put their 30 mm guns to maximum use. One flight of A-10s, flown by Capt Eric Salmonson and 1st Lt John Marks, destroyed 23 tanks. An Iraqi armored column surrendered to another flight of A-10s.[14]

Poised to attack directly into Kuwait, the Marine units were worried about the massed Iraqi artillery. As they conducted their breaching operations, they had forward air controllers and fighter aircraft monitoring the artillery counterbattery radar frequencies. Whenever somebody reported incoming rounds, the radars could quickly determine the location of the firing guns. The controllers would pass the coordinates to a forward air controller overhead, who would then direct an air strike against it. According to the commander of the 3d Marine Air Wing (3 MAW), Maj Gen Royal Moore, the system worked amazingly well.[15]

As the marines breached the Iraqi lines of trenches and wire, fighter aircraft were constantly overhead. The enemy was well supplied with air defense weapons. A flight of two F-18s transiting to hit targets near Kuwait City were both struck by heat-seeking missiles. Each aircraft lost an engine but was otherwise okay. The pilots flew them home to their base at Shaikh Isa, Bahrain.[16]

Almost all breaching operations were successful, and the Marine task forces and Tiger Brigade moved deep into Kuwait ahead of schedule. Watching the battle develop, General Schwarzkopf was pleased at the weak response of the Iraqis, but as the marines surged forward ahead of schedule, he became alarmed that they were becoming overexposed. Sensing that their flanks were unprotected, he called Lieutenant General Khalid and asked him if he could have the Egyptian and Syrian forces to the left of the marines move up their attack. The Arab forces refused, citing weather problems. General Schwarzkopf then called the commander the of 3d Army, Lieutenant General Yeosock, and directed him to begin the attack of the VII Corps early so that the forces fleeing Kuwait could not escape into southern Iraq.[17]

Caught by surprise, VII Corps nevertheless launched its attack that afternoon at 1430L when the 1st Mechanized Infantry Division unleashed its artillery. In the next 30 minutes the massed battalions fired 6,136 rounds and 414 MLRS (multiple-launch rocket system) rockets against the stunned enemy forces. Then, as the artillery guns reduced their rate of fire, combat engineers began cutting lanes through the enemy berms and lines of wire. The largest concentration of armored power since World War II was about to enter Iraq en route to Kuwait.[18]

25 February

The Iraqis appeared to have finally recognized the magnitude of the coalition force coming at them. Several of the Republican Guard and regular army armored units were ordered to establish screens behind which its forces could evacuate Kuwait. As the Iraqi forces retreated, they set fire to more of the oil wellheads. The raging fires produced billowing clouds of thick smoke. A weather front moved in with lowering ceilings and reduced visibility. The combination made it challenging for the aircrews to work and forced them to even lower altitudes where enemy gunners were waiting.

On the ground in Kuwait, the Iraqi III Corps counterattacked the 1st Marine Division as it came out of the barriers. Just after sunrise, two Iraqi mechanized brigades slammed into the right flank of the division. They appeared out of the Burgan oil field. The marines struck back with armor, TOW (tube-launched, optically tracked, wire-guided) missiles, artillery, and waves of air support. The battle was intense, and at one point, the Iraqis maneuvered to within 400 meters of the division command post.[19] Underneath the continuous air support, Cobra gunships stayed online and continuously launched more TOW missiles at any enemy vehicles that were not burning.

Another enemy force hit Task Force Papa Bear, the second element of the division to breach the enemy lines. The battle there was just as confused, and at one point fighting extended almost into the regimental command post before the Iraqis were repulsed. Flights of fighter aircraft were constantly overhead. The crew of an OV-10, flying in support of the division, spotted Iraqi forces regrouping on the east side of the Burgan oil field and was able to bring in AV-8s and Cobra gunships to destroy several vehicles and scatter the force.[20]

Approaching the Al Jaber Airfield in Kuwait, Task Force Ripper of the 1st Marine Division called for massive air support. A-10s, F-18s, and AV-8s answered the call, and Marine forward air controllers in OV-10s directed flight after flight of fighters and AH-1 Cobra gunships against the hapless enemy troops and tanks.[21]

(No mission number) – Jump 42/AV-8, USMC

That morning, as bad weather lingered over the Kuwaiti battlefields, Capt Scott Walsh launched on the wing of Maj Dan Peters out of their base at King Abdul Aziz Air Base, Saudi Arabia.[22] Their first sortie took them to a large enemy column opposing the Marine 2d Division. Recovering at the forward operating location at Tanjib, they refueled and rearmed their aircraft and took off again to fly another mission in support of their fellow marines. Entering the target area Walsh got separated from his flight lead. Staying below the clouds, he attempted to rejoin on another flight of Harriers. As he maneuvered, his aircraft was hit by a heat-seeking missile. The missile impacted the right rear exhaust and severely damaged the aircraft.[23]

Initially angered at having been hit, Walsh quickly surveyed the damage. "The blast blew a lot of the right flaps off, put several holes in the wing, and set fire to the fuel in the wing tanks," he remembered. The fast forward air controller (FASTFAC) working with him, Combat 13, joined up on him to inspect his aircraft. Not impressed with the flames streaking behind Walsh's aircraft, he suggested that Walsh eject.

Walsh observed the savage ground battle raging below him and decided to stay with the aircraft as long as possible. He jettisoned all of his ordnance to lighten his aircraft. The FASTFAC suggested that he land at the Al Jaber Airfield that was in the process of being liberated that morning. As the squadron intelligence officer, Walsh knew that the airfield was scheduled to be seized that morning by the 1st Battalion of the 7th Marines, part of Task Force Ripper. He concurred with the FASTFAC and set course for the airfield. As he prepared to land there, Captain Walsh realized that the airfield was not, in fact, under friendly control. Furthermore, his landing gear would not lower. He quickly considered making a vertical landing, but his nozzle control was not responding properly, and the aircraft was barely controllable.[24] All of these factors ruled out landing at Al Jaber. Walsh shoved his throttle forward and overflew the airfield. He would try to make it to friendly lines before ejecting.

As he turned to head south, his hydraulic system pressure slowly depleted and his flight controls froze. No longer able to control the aircraft, he was forced to eject. As the F-18 capped him from above, he floated to the ground in sight of forward elements of the task force. Immediately, the pilot in Combat 13 called the Marine TACC and reported that Jump 42 had been shot down and the pilot was alive on the ground. They quickly began to form a helicopter task force to get him out, but commanders on the ground had also watched Walsh go down. Almost simultaneously, calls were made to the forward elements of the task force to send out patrols to recover the young pilot.

Landing near an abandoned Iraqi bunker just west of the runways, Walsh oriented himself and quickly took shelter in a trench. He took out his pistol and radio and then called the aircraft above to let them know that he was okay. Once they acknowledged, he started moving south. Within a few minutes he encountered one of the Marine infantry teams who had been dispatched by the task force. They rescued him and took him back to the rear. There he gave their intelligence section a quick briefing on what he had seen, then got on a helicopter that took him back to his base.[25]

I084 – Pepper 77/OV-10, USMC

Working a few miles out from the 1st Marine Division, another OV-10 forward air controller was looking for remnants of the Iraqi force that had attacked that morning.[26] Maj Joseph Small and Capt David Spellacy were shot down by a heat-seeking missile five miles southwest of Mina Abd Allah, Kuwait, while supporting the drive of the 1st Marine Division. A ground forward air controller, Shotgun 212, observed the explosion and immediately tried to mount a rescue with ground forces. The pilot of an AV-8 working with Pepper 77 reported that he watched Iraqi AAA guns bracket the small aircraft and saw the launch of the missile. Major Small described what happened. "I was in a shallow right hand turn when we were hit on the right wing by a shoulder launched missile. . . . There was a loud explosion. The aircraft immediately went out of control."[27]

Small fought to control his aircraft, but it was severely wounded and could not respond. He ejected. After his parachute opened, he took out his survival radio and made an emergency call.

Watching the OV-10 spiral down, the pilot of the AV-8 saw the parachute and listened for an emergency beeper, but he never heard Small's call. Major Small landed among enemy troops and was immediately captured. Captain Spellacy had been killed in the explosion.[28]

The loss of the OV-10 and AV-8 in the area of Al Jaber Airfield—both to heat-seeking missiles—alarmed the USMC commanders in the area. Fearing more losses, they cancelled a helicopter-borne assault into Al Jaber. The airfield would be taken the old-fashioned way—by ground attack.[29] The concern of the commanders was well placed because 71 percent of all coalition aircraft shot down in the conflict were downed by Iraqi heat-seeking missiles.

In support of the overall movement of the marines, US ships in the Gulf launched 10 helicopters to stage a fake amphibious attack along the Kuwaiti coast. Flying at less than 200 feet, the helicopters carried special electronic emitters that replicated a much larger force determined to land in the vicinity of the port of As Shuaybah. At the last minute they pulled up and turned away. Iraqi forces placed along the coast to repel such an attack immediately responded with AAA and two Silkworm surface-to-surface missiles aimed at the coalition fleet. One of the missiles fell harmlessly into the Gulf. The other headed toward the battleship USS *Wisconsin* but was engaged and destroyed by British gunners on HMS *Gloucester*.

Ashore, the advance of the Marine forces continued unabated as they progressed to the western beaches of the Khalij Al Kuwayt, the bay that protrudes into the middle of Kuwait. There the Tiger Brigade attached to the 2d Marine Division bottled up the remaining Iraqis from the southern part of the country at the city of Al Jahra. With the main highway blocked, a huge traffic jam formed. This was detected by several intelligence sources and reported initially to the I MEF (Marine Expeditionary Force) commander, Lt Gen Walter Boomer. He ordered his A-6s, AV-8s, and F-18s of the 3 MAW to attack it.

Other intelligence sources also picked up this movement. The Marine aircraft were soon joined by Air Force and Navy fighters, as the TACC began diverting ever more flights to the growing target. Communications intercepts reported that Baghdad had ordered a general retreat of its forces out of Kuwait.

Headquarters elements were to lead so forces could be quickly reconstituted as they moved north of the Euphrates River.[30]

At the TACC, General Glosson personally called the 4th Fighter Wing. He wanted to get its F-15Es into this battle. Speaking directly with the vice wing commander, he ordered him to get 12 aircraft airborne as soon as possible and up to the Basra area. The aircraft had just returned from night missions and needed to be rearmed and refueled. The crews were beyond their legal limits for crew rest. Glosson waived that and told them to fly, noting to himself that the poor weather and oil fires would force them to work dangerously low. Privately, he expected as many as three would be shot down.[31]

The American press learned of the jammed highway and named it the "highway of death." In fact, few people were actually killed there, but the spectacle of the uncounted wrecked and burned vehicles became a metaphor for the fate of the Iraqi army.[32]

To the east, the 1st Marine Division steadily advanced toward the Kuwait International Airport and then Kuwait City itself. After destroying another Iraqi armored force, they secured the outskirts of the city. Then they coordinated with the Saudi and Qatari forces of JFC-E and the Egyptian forces in JFC-N to allow their units to pass through the Marine lines and enter Kuwait City in the van of coalition forces.[33]

To the west, VII Corps had gotten off to a slow start and was not through the barriers at sunrise. After some heated phone calls, the pace of VII Corps picked up as their heavy units pressed through the gates cut in the Iraqi defenses. Enemy forces began to react. A large force was detected advancing west toward the town of Al Busayyah. The leading element of VII Corps was the 2d Armored Cavalry Regiment (ACR). They slammed into enemy forces near the town and, after calling in A-10s and other fighter aircraft, destroyed them.[34]

Following the 2d ACR was the 1st Armored Division. It assaulted the town and, after an all-night battle, secured it and its huge munitions storage area the next morning.[35]

On the right the other divisions of the corps were in ever-increasing contact with enemy units. Some surrendered. Many did not. To the right of the VII Corps, the 1st Cavalry Division was the strategic reserve for CENTCOM. It was initially assigned the task of staging a diversionary attack up the Wadi al-Batin, the avenue of attack expected by the Iraqis.

(No mission number) AH-64

This aircraft was assigned to the 1st/227th Attack Helicopter Battalion of the 1st Cavalry Division.[36] While covering the division's initial attack into the tri-border area, battalion helicopters came in contact with enemy forces. In the melee, an Apache was hit by a SAM and crashed. The aircraft hit the ground hard, but the crew of C Company commander, CPT Mike Klingele, and CW4 Mike Butler was able to escape the aircraft. Enemy forces increased their fire. Immediately the second aircraft in the flight landed near the wreckage, and both pilots sprinted to it. They were wearing special harnesses with attachments that could be safely connected to the underwings of the Apache for immediate extraction. Quickly snapping in, they signaled for the pilot to lift them out of danger. With the two men securely attached, the Apache lifted them out to a safer area where they transferred to an OH-58C.

Slightly wounded, Butler was transferred to the battalion command and control helicopter. Captain Klingele stayed on the OH-58 and resumed directing his company as it reattacked the enemy forces.[37]

I085 – Health 67/Dalphin, RSAF

The aircraft was second in a flight of two helicopters flying routine logistics missions in the rear areas.[38] The flight entered some thick fog. The pilot of Health 67 broke formation and tried to turn out of the weather, crashing 30 miles west of Dhahran. A passing motorist reported it to RSAF headquarters. They reported it to the JRCC but took the mission themselves.[39] Compared to the rest of the TACC, things were relatively quiet in the JRCC.

That evening VII Corps headquarters received a new order. They were to pivot to the east and destroy the Republican Guard over the next two days. CENTCOM intelligence had detected that Iraqi units were beginning to pull out of Kuwait. General Schwarzkopf wanted his heavy corps to swing 90 degrees to the right and trap them before they got away.[40]

26 February

In the early morning hours, Marine helicopters lifted off the amphibious assault ships USS *Nassau, Guam,* and *Iwo Jima* to conduct feint operations against Iraqi units located on islands at the northern end of the Gulf. Escorted by AV-8s off the same ships and A-6s from larger aircraft carriers, they attacked enemy units to support the feint. While in the area, the pilots reported an increasing flow of vehicles moving north out of Kuwait.[41]

Far to the west, elements of the XVIII Airborne Corps had moved into the Euphrates River valley and were turning east to attack along Highway 8. With massive air support, the 24th Infantry Division seized the major Iraqi airfields at Tallil and Jalibah. Its powerful armored forces then continued east as the left flank of the entire operation. To their right the 101st Airborne Division was moving its forces forward and preparing for large air assault operations toward Basra.

To the south the VII Corps with its four heavy divisions and an armored cavalry regiment was beginning to turn east and attack into the heart of the Iraqi force. Throughout the day and night its forces would destroy, piecemeal, several Iraqi armor and infantry divisions in sharp battles with heavy support from A-10s, F-16s, and AC-130s.[42]

That afternoon General Schwarzkopf received a call from General Powell, who had become concerned at the tone of world press since the coalition forces had begun destroying the Iraqi army. The JCS chairman told the CENTCOM commander that a cease-fire was being discussed in many quarters. Schwarzkopf called his commanders and told them to speed up their attacks against the Republican Guard and begin planning ways to bring the combat to an end.[43]

As the enemy forces fled from southern Kuwait, the units were crowded into an ever-shrinking area around Al Basra. Both air and ground commanders wanted to get at them, but there were some complications. The weather was terrible. Low clouds and billowing thunderstorms continued to hang over the area, and the smoke from the oil fires just made it worse.

The enemy forces may have been retreating, but they were maintaining some unit cohesion and were still heavily armed and willing to fight back. To deconflict the air and ground efforts, Army commanders moved the fire support coordination line (FSCL) forward to north of the Euphrates River.

This was an important issue, because the FSCL is of great doctrinal importance. Beyond the line—away from friendly ground units—US attack aircraft could bomb targets free of any concerns that they might interfere with friendly ground operations. Inside of that line, though, any air strikes had to be under the control of Air Force air liaison officers or forward air controllers. These controllers were assigned directly to the Army ground units they supported and were positioned to provide coordination so there would be no interference, or worse, fratricide.

As the ground battle swept forward, senior Army commanders changed the FSCL often. This caused continuous friction with the ongoing air attacks against the fleeing enemy units. The air commanders were concerned and increasingly focused their attention on the problem. It also caused a lot of confusion among the aircrews flying the missions.[44] One of the airborne forward air controllers, CPT Scott Fitzsimmons recalled that it was difficult to stay up with the constantly changing FSCL amid the demands placed on him by the constantly changing battle situation below and the ever-dangerous Iraqi air defenses.[45]

27 February

In the morning an OA-10 working with the 1st Armored Division was hit by an Iraqi heat-seeking missile. The blast severely damaged the hydraulic systems and made the aircraft difficult to fly. The young pilot, 1LT Pat Olson, was able to keep the aircraft flying by switching the flight controls to the manual reversion system. This allowed him to control the aircraft—barely. The backup system was designed to give a pilot just enough control to fly out of the battle area and land. Undeterred, the young pilot decided to attempt a landing at the FOL at KKMC. He set up for a straight-in landing, but about a half mile from the runway, he lost control of the damaged aircraft. It rolled to the left and crashed in a huge ball of flame. He did not eject. The explosion scattered aircraft parts in all directions with many landing on six A-10s sitting at the end of the runway waiting to take off.⁴⁶ Several of the aircraft were damaged and could not fly their missions.

A few minutes later, another forward air controller, CPT Scott Fitzsimmons, took off to replace Olson over the 1st Armored Division as it moved east with the VII Corps and mauled the Republican Guard.

I086/M019 – Magic 14/AV-8, USMC

The pilot, Capt Reginald Underwood, was the wingman in a flight of two Harriers that launched from the USS *Nassau* (LHA-4) in the Persian Gulf.⁴⁷ Under the control of a USMC FASTFAC in an F-18, they were directed to bomb enemy vehicles fleeing Kuwait along the highway north to Basra. In the melee, enemy forces launched numerous heat-seeking missiles. Underwood's aircraft was hit by at least one of them. "I'm hit! I can't control it!" he screamed on the radio. When his aircraft hit the ground, the FASTFAC immediately assumed on-scene commander duties and initiated a rescue effort. JRCC passed the mission to SOCCENT. They launched two MH-60s and an MH-47 that held at the border waiting for report of a position and voice contact. The JRCC diverted A-10s to support them, but Captain Underwood had been killed.⁴⁸

To the northwest, the XVIII Airborne and VII Corps had turned east and were attacking the Republican Guard. All of their aviation units were heavily engaged in supporting their parent units.

(No mission number) – UH-1, 507th Medical Battalion

The helicopter attached to the VII Corps was hit by enemy fire while engaged in a medical evacuation (MEDEVAC) mission. AAA damaged the tail rotor causing the aircraft to go into an uncontrollable spin. It crashed killing three crew members and injuring the fourth. The Army RCC directed the recovery operation.⁴⁹

(No mission number) – UH-60, 1st Infantry Division

Elements of the XVIII Corps discovered the crash of this aircraft that had been shot down on a ferry mission. All six persons on board had been killed. The Army RCC again directed the recovery operation.⁵⁰

1087/M020 – Mutt 41/F-16, USAF, and Bengal 15/UH-60, US Army

The XVIII Airborne Corps was making plans to airlift a brigade of the 101st Airborne Division to a position near Basra to trap remnants of the Republican Guard units as they tried to flee. Everybody was trying to get at the panicked enemy troops. A flight of four F-16s from the 10th TFS of the 363d TFW was launched to provide CAS for the advancing US Army forces.⁵¹ They were not immediately needed and diverted to the airborne controller for a target. They were instructed to hit fleeing Iraqi units along the Basra road.

About 15 miles away, CPT Scott Fitzsimmons in his OA-10 had arrived in his assigned area. He recalled, "The weather was real bad. I went in and talked to Warmonger—the Air Support Operations Center of the VII Corps. They pushed me forward to an Alpha [holding/contact] point. I talked to the [1st Armored] Division ALO. He gave me a tasking . . . and pushed me forward to Chowder 21 who was the ALO for the 2d Brigade."⁵²

Breaking out below the weather at about 3,000 feet, the young pilot could see a line of M-1 tanks and M-2 Bradley fighting vehicles several miles long, running north-to-south, and all of them were slowly moving east. He could see that they were firing, and in front of them were burning Iraqi tanks, personnel carriers, and trucks. Some of the Iraqi T-72 tanks were firing in return, but their shots were bouncing off the M-1s.

The visibility was about three miles with rain showers. The brigade ALO directed him to attack a tank position several kilometers in front of the M-1s. Fitzsimmons rendezvoused with a flight of two A-10s. They were carrying Maverick missiles. He marked the target for them with a smoke rocket. When the ground forward air controller with the lead battalion confirmed that he had marked the enemy tanks, he cleared his A-10s to attack them. The two A-10s made several passes and destroyed the designated vehicles with missiles and 30 mm cannons. They were low on gas and had to depart for home.⁵³

High above and a few miles to the north, the four F-16s of Mutt flight arrived over the assigned coordinates. The flight lead, Capt Bill Andrews, flying his 35th combat mission, split his flight and took his number two man down below the clouds and smoke to look for targets. Scanning through the smoke and haze, Andrews did not see anything that appeared lucrative. As he added power to climb back above the clouds, his aircraft was wracked by a violent explosion. Andrews saw warning lights illuminate in the cockpit. He could see a bright yellow glow in his rearview mirror and knew that his aircraft was on fire. As it pitched over and started to dive at the ground, he ejected.[54]

Descending in his parachute, Andrews took out his emergency radio and made several panicked calls as he drifted through the smoke of the raging battle below. Captain Fitzsimmons heard him loud and clear. His response was automatic:

> I stuck my nose up and responded. I didn't know how close Andrews was. I could hear him well on the radio. . . . Somebody came on the radio and gave a bulls eye call, "Randolph 200 degrees for whatever it was." I whipped out my smart pack and found Randolph, punched in the coordinates in the INS and according to the INS, I was pretty close. So I just started heading north thinking that I could help him out. What I did not know was that the line of tanks that I was working over only extended about 3 or 4 miles to the north. He was about 12 miles north. As the tanks moved east they weren't, I don't know if they were securing their flank or what, but as I went more than about 4 miles north, I was back in enemy territory, but I did not know it. . . . I was at about 2,500 to 3,000 feet. I got hold of him [Andrews]. He was in his chute, near "the factory," which was an F-16 geo-reference point. It was unknown to me. He said that he was coming down just north, just west of the factory.[55]

Hanging in his parachute, Andrews winced as large-caliber rounds from the numerous AAA guns flew past him. As he neared the ground, that firing stopped, only to be replaced by the whizzing of rifle bullets.

Andrews hit the ground hard and broke his right leg. He released his parachute and watched several Iraqi soldiers run towards him. Several were firing their AK-47s. He laid down his radio and raised his hands. They stopped firing. Fitzsimmons flew overhead, unaware of Andrews' situation.

When the soldiers were about 30 feet away, Andrews saw a SAM being fired behind them. It was tracking the OA-10 flown by Fitzsimmons. He quickly grabbed his radio and called for the OA-10 pilot to break and drop flares.

Above, Fitzsimmons instinctively reacted. He remembered of Andrews, "He made a radio call, 'BREAK, ZSUs, MISSILES!' I heard that so I just pushed on the stick and almost instantaneously, this wall of fire goes over my canopy, I am guessing from the ZSU. And then I started hammering on the flares and he made some kind of 'BREAK LEFT, FLARES, FLARES' call. He kept hollering on the radio. . . . And when I looked back two IR [infrared heat-seeking] surface-to-air missiles had exploded behind me, biting off on the flares."[56]

When the soldiers saw what Andrews had done, they began firing at him and charged. He dropped the radio and again raised his hands. They shot his radio to pieces and then did the same to his helmet, raft, and other assorted equipment. They then stripped him of the rest of his gear, threw him in the back of a pickup truck, and took him to a headquarters.[57]

Reacting as trained and unaware of the events below, Fitzsimmons started the process of notification. He reported Andrews' loss and a quick situation report to AWACS, who forwarded the information to the TACC. Then, low on gas, Captain Fitzsimmons had to depart the area.

Fitzsimmons was able to note Andrews' general location. Understanding the overall tactical situation, he said, "I could not pull off a CSAR that fast. Even if I had been able to get my eyes on where he was, I do not think that there was anything that I could have done—we could have had a hundred airplanes and I am not sure we could have done anything. [A rescue effort] was probably not a good idea, especially since he was captured and they—I am sure that I told the [airborne controller] as I was heading out that I was pretty sure that they [Iraqis] had got him because he was off the radio."[58] Another OA-10 forward air controller in the area, CPT Gerry Stophel, recorded the entire event on a small cassette tape recorder he carried in his cockpit.[59]

The notification of Andrews' downing came in from AWACS to the TACC and was quickly brought back to the JRCC. Plotting the survivor's location, Lt Col Joe Hampton tasked the mission to SOCCENT since it was beyond the FLOT. The JRCC intelligence cell also quickly looked at the situation. They could see that the survivor was down in an extremely dangerous location.[60]

Word of the downing of the F-16 in the midst of the ground battle rippled through the TACC. Maj Lornie Palmgren, a SOCCENT liaison officer working the day shift there, remembered hearing about it. He noted Andrews' position on the map, then walked over to the JRCC intelligence cell. Scanning the locations of all the Republican Guard units in his immediate vicinity, he recalled that, "It took me about 30 seconds to make a decision."

He said to Joe Hampton and the director of the TACC, Col Al Doman, "We're not sending two crews out there. That is a high threat area. That is absolutely ridiculous. Let's not do this . . . we can't handle it. It was daytime. If it had been a nighttime thing we might have taken another look at it."[61]

At SOCCENT the intelligence officer on duty, Capt Jim Blackwood, took the call. He and his enlisted assistant quickly plotted the downed airman's location on the SOCCENT threat board. Like the officers in the JRCC, they could immediately see that he was down in the midst of the Al Faw Republican Guard Division and quickly came to the conclusion that any attempted helicopter rescue in daytime would be far too risky.

Colonels Gray and Johnson were both in the SOCCENT command center when the call came in. Captain Blackwood briefed them on the situation. Colonel Gray recalled, "I was sitting there in the room with Jesse Johnson and we're looking at the intel picture along with our intel people and we said, 'Uh-uh! This is bad news. We're not going to do this. There is just no way.' It wasn't a unilateral decision. That was a joint decision among all of us. 'We're not doing that. As much as we hate it we just can't go in there. . . . We'd really like to get that guy, but it was just going to be bad news.'"[62]

Collectively, they came to the conclusion that first of all, the risk of losing a rescue helicopter was much too high. Also, the probability of rescuing the young captain, especially in daylight, was too low.

Perhaps a night effort was a possibility, but they refused to fly the mission in daylight and so notified Colonel Hampton.[63] Hampton took the call in the JRCC. The refusal led to a heated discussion between several JRCC personnel. Several questioned the decision by SOCCENT since helicopters had launched just a few hours prior for Magic 14, who had gone down about 30 miles southeast of Andrews' location. One young USAF officer went over to talk to a senior ranking USAF officer in the TACC. The senior officer followed the younger man back to the JRCC area and overheard the ongoing conversation.[64]

As one who had been shot down and rescued himself in Southeast Asia, the senior officer understood the importance of rescue and was not happy with the answer, but there were several problems ongoing at the time. A larger one had to do with the rapid movement of the massive Army units, and the senior officer had been closely monitoring their movement as they swept towards Basra.

The fast advance of the Army units dictated the constant changing of the FSCL. The frequent movement of the line eastward as the Army VII and XVIII Corps pressed forward was directly impacting the efforts of the coalition aerial forces to attack the fleeing Iraqi forces.[65]

The XVIII Corps had been augmented with a strong force of attack helicopters. It was pushing its lead division, the 101st Air Assault Division, to launch its AH-64 Apaches against the same Iraqi units. They needed to use the same airspace as the Air Force, Navy, and Marine flights above. To deconflict these operations and prevent fratricide, the FSCL had to be properly placed. Capt Bill Andrews went down in the middle of this mess.

The senior USAF officer had been working this issue directly with the Army battlefield coordination element (BCE) in the TACC. After receiving the news about Mutt 41 and the refusal of SOCCENT to launch a helicopter, he walked over to the BCE. He pointed to Andrews' position on the map. Hampton remembered what happened next:

> [The senior USAF officer] said, "You got any choppers up there? Why don't you pick the guy up?" And so they launched an Army chopper. . . . I was in the back of the room, and one of the guys said, "Hey, you had better go up and find out what is going on. I think we have got something happening." I went up there [to the BCE] and [the senior Air Force officer] told me what he did. [He said], "I talked to the Army guys. Is there any way to recall them?" I said no.[66]

Others were attempting to recover Andrews as well. When word of the loss of Mutt 41 came into the operations center at KKMC, some of the F-16 guys remembered the effort that had been made for Scott Thomas, Benji 53, and on their own, decided to try that approach again. A few went across the field to visit the Army flyers of the 3d/160th SOAR. One of the Army pilots present, CW2 Steve Rogers recalled,

> They came across the field with the ISOPREP (individual identification information for Captain Andrews) info and said "You have to go get our buddy." Lieutenant Colonel Daily, the 3d/160th commander said, "Hey I know how you feel, but I can't launch on what you are doing here. We will send it up and see what happens." This information then got passed up to [SOCCENT]. They turned around and said "Hey, that is good information." They then called us back and told us to start planning to go recover him. So we started planning a night mission. Our call sign was Asp. We planned a two-ship mission. I sat down at the threat board with the S-2 [intelligence officer] and he seemed to be well versed in the threat. We drew a straight line from our location to where he was and realized that there was just no way. We had to cross the ground battle. There was just so much air defense stuff going directly to him or even just trying to scoot around. It was just going to be impossible. So I just said that the only way to do this is to go out over the water, go up by the Shott, turn around and come in from the north. We had the fuel to do it that way.[67]

Back in the TACC, the Army controllers in the BCE called the "order" forward to the XVIII Corps. There the commander, Lieutenant General Luck, was skeptical that it was viable and wanted as much information on it as possible before committing to the mission. He was unaware that the mission was already being called forward to the aviation brigade of the 101st Division.

At the time the 101st had several aviation battalions forward at a location called objective "Tim" that had been seized by the 3d Brigade of the division just a few hours prior. "Tim" had been turned into a forward operating base for the aviation elements and was now called "Viper." Here the attack battalions were refueling, rearming, and launching to attack Iraqi units fleeing north towards Basra. The situation was chaotic, and the attack crews were being launched with minimum intelligence or clear objective other than to attack and destroy targets of opportunity.[68] Hearing that XVIII Corp assets were going to attempt a rescue, Colonel Gray at SOCCENT called down to the 3d/160th and told those guys to "turn off their efforts" because regular Army units were going to try and get him.[69]

One of the first units into Viper was the 1st Battalion of the 101st Aviation Regiment (1/101) commanded by LTC Dick Cody. Its companies were assigned missions upon arrival. CPT

Photo courtesy of Bill Bryan

LTC Bill Bryan commanded the 2d Battalion of the 229th Aviation Regiment (2/229), one of the most experienced Apache units in Desert Storm. As a young lieutenant, he had been assigned to F Troop/8th Cavalry in Vietnam in 1972 when one of their helicopters, Blue Ghost 39, was shot down and the crew killed or captured trying to rescue Bat 21 Bravo. It was a tragic precursor to Bengal 15.

Doug Gabram, the B Company commander, was ordered to take his Apaches and attack to the northeast along the causeway over a large bog called Al Hammar. He was in the target area when he heard Andrews' emergency calls. The calls were garbled. Reacting instinctively, he quickly checked his flight to determine that none of his aircraft were down. He then tried to determine Andrews' location based on what was being said to the forward air controller, Nail 51. When the survivor stopped talking on the radio, Gabram felt that it was just too risky to fly east to search for him.[70]

Another battalion that went into Viper was the 2d Battalion of the 229th Aviation Regiment (2/229). A separate unit originally located at Fort Rucker, Alabama, it had traditionally been a training/conversion unit for newly assigned Apache pilots, but as Desert Shield had grown, it was brought up to wartime equipment and personnel levels and deployed to the Gulf. There, it was assigned to the XVIII Corps and attached to the 101st. It had 18 AH-64 Apaches, three UH-60 Black Hawks, and some of the most experienced Apache and Black Hawk pilots in the Army. Several had logged extensive combat time in Vietnam. One of them was the commander, LTC Bill Bryan.

When his lead elements flew into Viper, Bryan met with BG Hugh Shelton, assistant division commander of the 101st, and the 101st Aviation Brigade commander. As his helicopters were being refueled and rearmed, Bryan received his instructions. Little intelligence was available for his unit's mission. He was told to "fly due north, hit the Euphrates River, fly east along the river until you hit the causeway." There, the 2/229 was to attack fleeing Iraqi vehicles in a designated engagement area called "Thomas." Bryan was urged to move quickly since it appeared that the war would soon end and his commanders wanted to destroy as many Iraqi forces as possible before any cease-fire.[71]

With his mission guidance, Bryan directed his three attack companies to sequentially strike along the causeway. He would lead the effort with C Company. B Company, with the unit operations officer, MAJ Mike Rusho, and his assistant, CPT Dave Maxwell, would be second in one of the other UH-60s; and A Company would be third. Briefing complete, Bryan's commanders saluted and proceeded to carry out their missions.[72]

The Apaches that Bryan's crews flew were specifically designed to destroy enemy armor. The Black Hawks were assigned to the unit for utility support. As the Apaches were striking the enemy units, the Black Hawks were moving unit equipment forward to Viper. They were also launching behind the Apaches for possible intra-unit rescue duty.

The order to rescue Andrews was passed down through the 101st and to a senior officer at the aviation brigade.[73] Initially, he offered the mission to Colonel Cody and his battalion of Apaches. Cody looked at the map. Quickly realizing where the survivor was, he concluded that the mission was far too dangerous and recommended that nobody fly in there.[74] The brigade senior officer also knew that the 2d/229th was at Viper. He called for the commander, Lieutenant Colonel Bryan, but was told that he, his operations officer, and assistant were out flying missions. The operations officer, Major Rusho, had left another young assistant, 1LT Al Flood, behind to organize the rearming and refueling operations and to push the gunships forward. Flood had just arrived in another Black Hawk flown by CW4 Philip Garvey and CW3 Robert Godfrey. Their call sign was Bengal 15. They dropped him off and repositioned to the refueling pit to take on gas. Lieutenant Flood related what happened:

> I heard someone call me on the radio and it was the [aviation] brigade TOC [tactical operations center]. They were trying to contact our elements forward but could not, so the brigade TOC contacted me, asked where I was. I said, "I'm about 200 meters south."
>
> He said, "Roger, come over to the brigade TOC, I need to talk to you." So I . . . walked over there. I talked to the Brigade [senior officer] and that's where he briefed me on the situation.
>
> In a nutshell, what he told me was that we had a situation up north, we were receiving reports of an F-16 pilot who had been shot down over Republican Guard units that were near where my battalion elements were in contact with them. He asked me what we had here in the rear, i.e., at FOB Viper. I told him we had one company which was A Company, 2d/229th that was ready to launch within 30 minutes. We had some OH-58s.
>
> Then he said, "Do you have any CSAR assets back here?"
>
> I said, "We have Bengal 15 on the ground. He just came out of hot refuel."
>
> He said, "Really?"
>
> I said, "Yes sir."
>
> He said, "I tell you what. Go get the crew and tell them to talk to me. I need to talk to them about this."[75]

Lieutenant Flood then ran to get the two pilots of Bengal 15. He described what happened when he reached them:

> I said, "Listen, the [senior officer] needs to talk with you. We have a rescue mission."
>
> So they walked over to the brigade TAC and he briefed them on what was going on up north. What he asked them was "Can you guys go and get him?" Well, the obvious answer was "Yeah!" They were the best UH-60 pilots that we had in the battalion. That was our mission—for the UH-60s inside 2d/229th, command and control, aerial resupply and rescue.
>
> About 5 or 10 minutes later, they were walking back towards the aircraft, and I asked Garvey and Godfrey what was up.
>
> They said, "Well, got an F-16 guy and he's been shot down over Guard units. We're going to go up there and get him."

I said, "Okay. What else?"

Garvey said, "What we want to do is we're going to get two of the AH-64s to give us an escort up there."

I said, "With the A Company guys, right?"

They basically said, "Yeah."

What I did then was I said, "Okay. You guys are going to crank right now, right?"

They said, "Yes. We're heading there now."

I said, "I will contact A Company and let them know what the plan is and I will also tell them to contact you as soon as they get ready so you can give them the details of the mission. Then you can launch off together."

I also knew that part of the plan was once they got up farther to the north, closer to where the actual shoot down had been, that the plan was that they were going to contact AWACS on guard and the AWACS would vector them in. They explained that to me as well but we were in a rush to try to get the mission off as fast as possible so I didn't pick their brains for every single detail. Bottom line was they had a mission, I knew the basic plan of it. Then I contacted A Company on my FM radio.[76]

The pilots of Bengal 15 prepared to take off. They had received only a generic intelligence briefing on the situation and no specific information on the particular area into which they would be flying. They had tactical maps of that section of the theater, but they had not been updated on the constantly changing battle situation.[77] In the quick discussion with the senior officer, he had also instructed them to contact their commander, Lieutenant Colonel Bryan as soon as they got airborne.[78]

The Apaches belonged to A Company of the unit and were being led by the company commander, CPT Mike Thome. Garvey quickly contacted Thome on the radio. Thome agreed to cancel his assigned deep-strike mission and provide escort for the rescue effort. He and Garvey rapidly put together a plan of action. The battalion had an intra-unit rescue plan if an Apache went down, but a rescue mission for someone not in the unit was unrehearsed and needed some thought.[79]

The senior officer had also told Garvey that the downed airman was wounded. Garvey asked that the unit flight surgeon, MAJ Rhonda Cornum, go with them to provide assistance. She had ridden in with them to Viper earlier and was nearby. Moments later three Army Pathfinders arrived to accompany them

The crew of Bengal 15 included, *left to right,* CW3 Gary Godfrey, CW3 Phil Garvey, SGT Dan Stamaris, and SGT William Butts.

for site security.⁸⁰ All were totally pumped up at the thought of performing a rescue. Major Cornum later wrote, "This was the real thing, combat search and rescue. There was no mission that we trained for that was more important, more exciting, or more dangerous. My heart beat faster, and my stomach tightened. This was it. We were doing it for real."⁸¹

When everybody was ready, Bengal 15 flight lifted off. Garvey contacted AWACS, and the controllers began giving them vectors to the survivor's location. The crew did not contact the battalion commander, Lieutenant Colonel Bryan, who had been listening to the F-16 pilot as he descended in his parachute. Bryan had a unit plan for such pickups and could have rendezvoused with them for a larger effort.

While these preparatory events were taking place at Viper, in the skies over Andrews' position, AWACS had started diverting strike flights for the CSAR. One of the first was a flight of four F-16s, also from the 10th TFS. It was led by Capt Mark Hebein. Checking in, the forward air controller, Nail 51, asked him to go down below the clouds and see if he could find the survivor or at least find his wreckage.

Hebein was carrying a load of CBU-87. He left his flight above the clouds and dropped down for a look. He described what he saw:

> I left my flight orbiting above. . . . I said okay, . . . I will make one more pass, this time, I will go down from east to west and then swing around out of the sun because it was just setting. I . . . broke out at about 3,500–4,000 feet. Immediately, I could see the crash site and burning wreckage, about a mile south of me. I said, "He is somewhere between there and the road," so I started swinging around and as I looked back I had a couple of SAMs coming at me. Probably SA-16s . . . so I rolled out and as I did, I saw immediately all around me red fire-hoses, ZSU 23-4 within 50 feet on all sides. I thought, "Jeez, I am just toast." So I threw it into afterburner and pulled back as hard as I could and closed my eyes because I just knew that I was going to get blown out of the sky. . . . About this time they [AWACS controllers] are asking me, "Were you able to see him because we got the helicopter going in," and I said, "No, the area is way too hot. Do not send a helicopter in right now." I tried to tell them that about three times and it was broken on the [secure] radio and I said the area was way too hot for any SAR right now. Do not head into that area, because I know with all that stuff coming up a helicopter would be toast. And they acknowledged it.[82]

Nail 51 was low on gas and passed the on-scene command to Pointer 75, an F-16. Holding above the clouds, Pointer continued to coordinate to get strike flights in to hit the guns and SAMs. Two F-18s from Marine Fighter/Attack Squadron 314 (VMFA-314) checked in with Pointer. Initially fragged for an interdiction mission, they, like so many others, were immediately diverted for CSAR support whenever required. Eager to help, the two Marine pilots, Capt "Otis" Day, and his wingman, Capt Paul Demers, set up for the strike as Pointer 75 gave them a cursory briefing on the situation as he knew it and the intense enemy antiaircraft threat.

The weather had worsened. Thunderstorms had popped up and were right over the area. The two F-18s were at 25,000 feet. The forward air controller was above them at 30,000 feet. Captain Day asked him for a more detailed briefing on the situation. The forward air controller could not provide one because he had not been below the clouds. He gave them the coordinates that he had for the survivor. Captain Day was aware of the threat. His intelligence officer had briefed them before the flight. Additionally, his radar warning receiver was barking at him constantly as new SAM sites tried to track their aircraft.

Captain Day took his flight down below the weather. They broke into the "clear" at about 8,000 feet. "Clear," though, was a relative term. In actuality, the visibility below the clouds was abysmal because of the oil fires and burning wreckage. As they cleared the clouds, the enemy gunners and SAM site operators spotted them and took them under constant fire. After frantically searching, Captain Day found what he thought was the wreckage. Pointer 75 told him that AWACS had said that a helicopter would be approaching from the west for a rescue attempt. Captain Day decided to attack anything that they could find extending west from the target area. Both men then dropped their CBUs and bombs on several targets. This was done without any apparent coordination with the elements of the 24th Infantry Division that were just west of the survivor's location and were, in fact, being overflown by Bengal 15. With their ordnance expended and now critically short on fuel, the two F-18s departed the area. Captain Day checked out with Pointer. He also informed him that the area was much too hot for a rescue attempt and recommended that they not try it.[83]

Back below the clouds, the flight of helicopters headed for the downed pilot. The Black Hawk was faster and moved into the lead. The Apache pilots had their aircraft at maximum power but steadily fell behind. They used the turns to close on the Black Hawk.

AWACS initially took the flight north. With no coordination, they were taken through the sector of the 24th Infantry Division. They overflew its lead unit, the 2d Battalion of the 7th, a mechanized infantry unit that was in contact with enemy forces. The visibility was poor due to all the oil fires and burning enemy vehicles.

The AWACS controllers turned them east to approach the survivor's location. As they neared the enemy lines, the flight began to take fire from enemy positions. Just then, the AWACS controller told them that they were "entering a red zone," and several larger-caliber weapons firing green tracers opened up on them. One of the Black Hawk pilots called, "Taking fire! Nine o'clock! Taking fire!" Simultaneously, several rounds slammed into Bengal 15. That was immediately followed by a plaintive

Photo courtesy of Bill Bryan

MAJ Rhonda Cornum, flight surgeon for the 2d/229, was captured when Bengal 15 was shot down. She is shown in an earlier photo with an unidentified Kuwaiti officer.

call from the stricken helicopter, "Don't put the rotor in the ground. Don't put the rotor in the ground, Phil!" The fatally wounded aircraft pitched forward and crashed into an earthen berm. Landing nose first, the aircraft dug into the soft sand and flipped over. Five on board were killed in the explosion. Within moments Iraqi soldiers were upon the surviving three, who were now prisoners.[84]

Killed:	CW4 Philip Garvey
	CW3 Robert Godfrey
	SFC William Butts
	SGT Roger Brelinski
	SSG Patbouvier Ortiz
Captured:	MAJ Rhonda Cornum
	SSG Daniel Stamaris
	SP4 Troy Dunlap

The enemy guns also found the Apaches. One of them was hit, with damage to the left engine nacelle and fuselage. Captain Thome watched the Black Hawk roll up in flames.[85]

Reviewing the situation quickly, Thome determined that the chance of survival was low and the risk to the Apaches and crews was too great, so he ordered the two aircraft out of the area.[86] The rescue had failed.

As word of the failure was passed along, SOCCENT called the 3d/160th and told them to again prepare for a night effort. CW2 Steve Rogers and his crew went back to work. Soon they got another call. Rogers said,

> Then they called us again and told us that it was a no go; that the area was just too hot. No way that they would approve a launch. So we shut down. Then we heard what had happened to them. Everything they had run into, we had plotted. That was the sad part. This crew had flown into the unknown. They gave them a mission and they had launched and flew right into bad guys and I just know that their threat map was just not as updated as ours. We had the best threat intelligence in the theater.[87]

Working their way through the growing thunderstorms, USMC pilots Capt Otis Day and Capt Paul Demers landed at the nearest airfield for gas, only to then learn that their efforts had been for naught, that the survivor had been captured and an Army helicopter had been shot down.[88] Likewise, Capt Mark Hebein and his flight recovered at KKMC and discovered the same thing. Captain Hebein remembered the moment: "I got back and debriefed and that's when I found out that a helicopter went in just a few minutes after that and had got shot down. I was [upset] because I had told them specifically. I was right there, and the area was way too hot to send a helicopter in, especially unescorted. . . . I am so [upset] that they sent a helicopter in there after I told them not to."[89]

Most shocked of all was the commander of the 2d/229th, LTC Bill Bryan. He had been a young lieutenant on his first assignment with F Troop of the 8th Cavalry in South Vietnam in 1972 when almost exactly the same thing had happened to a UH-1 crew. He remembered the loss of Bengal 15 clearly:

> As far as a Black Hawk going down, my plan was to go get the back-up Black Hawk and go after the first. . . . My experience from [Vietnam] had molded my feelings that you need to think long and hard about sending 4, 5, 6, 7 or more people to rescue someone who is possibly mortally injured, captured or dead. That may sound cold-blooded but at some point, someone has to say "no more." I was prepared to make that decision if I had to. That is some of the unwritten stuff that commanders get paid to do—but not necessarily trained.

Had I been there when this mission came up, the first thing I would have done would be to insist on knowing something about the ground situation. That information was non-existent because of the "rout." But what WAS known was that the Iraqis had shot down an F-16, the pilot had ejected and was talking on his radio going down . . . and the Iraqis were shooting at him in his chute. Knowing these facts (or lack of them) would have led me to conclude that the pilot had either survived and was now a POW or he was dead—murdered by Iraqi cowards who shot him before he hit the ground or shortly thereafter.

For me the next decision would have been easy. I would not put another 3–4 aircraft and 10–15 of America's sons and daughters at risk! Some would scoff at that but I remain firmly convinced that that rationale was valid then and is valid today.

CSAR is like any other military mission. It takes planning, coordination, and resources. Because we are dealing with life and death here and extreme urgency, many overlook the planning, coordination and resource requirement.[90]

In the JRCC, Maj John Steube shared the same thought when he remembered that, "People just wanted to get this guy. They stopped thinking about what the consequences were for the people who had to go in there. It cost a bunch of people their lives."[91]

The day was not over. With all units now in heavy action, more losses were almost inevitable.

I088 – Avenger 11/AH-64, US Army

The aircraft went down because of fuel starvation. The Army RCC dispatched a task force to recover the crew and aircraft, but the crew had been killed.[92]

I089 – Cyclops 47 and 48/2 OV-1s, US Army

The two aircraft could not safely land at Hafra Al Batin when the runway lights went out.[93] One aircraft crashed on the runway. The other crew did not have enough fuel to divert to another airfield and ejected a few miles from the base. The Army RCC dispatched helicopters. Two crew members were wounded.[94]

Throughout the night, the forces of VII and XVIII Corps continued to attack into the Iraqi units. The battles were fierce and one-sided. Iraqi units were destroyed piecemeal with minimal coalition casualties.

In Washington President Bush convened a meeting of his top security advisors to determine when to end the war. Ever solicitous of the opinions of his military leaders, he asked General Powell for his opinion. The general replied that General Schwarzkopf had indicated to him that he needed just a few more hours to achieve his assigned goals. After asking all the others present for their views and reviewing the political situation, he realized that the decision was his to make. With that the president determined that the cease-fire would take effect at 0500Z/0800L on the next day, 28 February.

That evening the president addressed the nation. Proudly, he announced: "Kuwait is liberated. Iraq's army is defeated. Our military objectives are met. . . . At midnight tonight, Eastern Standard Time, exactly 100 hours since ground operations commenced and six weeks since the start of DESERT STORM, all United States and coalition forces will suspend offensive combat operations. It is up to Iraq whether this suspension on the part of the coalition becomes a permanent cease-fire."[95]

But there were still a few hours to go. As the fighting continued, in the TACC, General Glosson got permission to hit a few more targets in the Baghdad area. Concerned that Saddam and his family might try to flee, he had aircraft at several airfields around the capital destroyed. He added some previously unidentified nuclear sites to the target list.

Getting the missions, the F-117 pilots were concerned. They had heard the president's speech. With the end near, nobody wanted to be the last casualty. Additionally, bad weather had moved into the Baghdad area. This would force them to make their bomb runs at much lower altitudes. So far, no F-117s had been lost.

That last night though, their luck almost changed. One F-117 had its bomb bay door stick open after bomb release. The interior metal was not shielded with the special ablative paint and presented a radar return. An SA-3 SAM suddenly locked on to the aircraft and launched a missile. Fortunately, the pilot was able to get the door closed before the missile could hit the aircraft. Another F-117 flying at lower altitude was almost hit by AAA.[96]

In the last hours of combat, four B-52s and an F-16 were damaged by Iraqi air defenses, but there were no more losses.

Avenger and Cyclops were the last recorded incidents prior to the cease-fire at 0800L on 28 February.[97]

As hostilities ended, rescue forces were still standing alert at several locations. The JRCC showed the following recovery assets available:

RSAF	–	20
US Navy	–	10 (SOCCENT)
US Army	–	12 (SOCCENT)
USAF	–	25 (SOCCENT)
FAF	–	2
RAF	–	1[98]

The next day, a task force from the 2d/7th Infantry Battalion of the 24th Infantry Division moved forward and secured the Bengal 15 crash site. They found the remains of the five soldiers killed and indications that the rest of the troops had been taken prisoner. Later that day, LTC Bill Bryan of the 2d/229th Attack Battalion was able to fly into the site and help recover the bodies of his men.[99]

Very quickly, the JRCC started working with CENTCOM to identify crash sites so that other teams could move in and secure them too—at least in those areas where coalition ground forces were in control. Three days later, General Schwarzkopf met with Iraqi military commanders at the Safwan Airfield to review the terms and conditions of the cease-fire. Item one was the return of all coalition POWs and bodies. There would be no repeat of the lingering national wounds of the MIAs from Southeast Asia. By 6 March, all coalition POWs were returned.

Then the forces started going home. Desert Storm was over.

Notes

1. Michael R. Gordon and Gen Bernard E. Trainor, *The Generals' War: The Inside Story of the Conflict in the Gulf* (New York: Little, Brown and Co., 1994), 245.
2. Ibid., 340.
3. Ibid.
4. Ibid., 348.
5. Ibid., 349.

6. History, 3d Marine Air Wing (3 MAW), Marine Attack Squadron, VMA-542, Washington, DC: Marine Corps Historical Center, 1 January–28 February 1991.

7. Dale B. Cooper, "Young Guns, Harrier Pilots in the Persian Gulf," *Soldier of Fortune*, May 1991, 52.

8. History, 3 MAW, 145.

9. CDR Patrick Sharrett et al., "GPS Performance: An Initial Assessment," *Navigation Journal* 39, no. 1 (Spring 1992): 411.

10. Michael R. Rip and James M. Hasik, *The Precision Revolution: GPS and the Future of Aerial Warfare* (Annapolis, MD: Naval Institute Press, 2002), chapter 5.

11. Williamson Murray with Wayne Thompson, *Air War in the Persian Gulf* (Baltimore, MD: Nautical and Aviation Publishing Company of America, 1996), 276.

12. "Desert Storm, Final Report to Congress: Conduct of the Persian Gulf War 1991," *Military History Magazine*, April 1992, 147.

13. Barry D. Smith, "Tim Bennett's War," *Air Force Magazine*, January 1993.

14. Murray with Thompson, *Air War in the Persian Gulf*, 286.

15. History, 3 MAW, 151.

16. Ibid., 152.

17. Gordon and Trainor, *Generals' War*, 362.

18. Ibid., 379.

19. Lt Col Charles H. Cureton, *U.S. Marines in the Persian Gulf, 1990–1991: With the 1st Marine Division in Desert Shield and Desert Storm* (Washington, DC: History and Museums Division HQ USMC, 1993), 93–95.

20. Ibid.

21. History, 3 MAW, 159.

22. This was not logged on the USCENTAF/JRCC Incident/Mission Log.

23. Col John Bioty Jr., interview by the author, 6 January 2000.

24. Dale B. Cooper, "Young Guns," 52.

25. History, VMA-542; History, 3 MAW, 161; and Capt John Walsh, interview by the author, 6 October 2000.

26. USCENTAF/JRCC Incident/Mission Log, DS box, Joint Personnel Recovery Agency, Fort Belvoir, VA, n.d. (hereafter USCENTAF/JRCC Incident/Mission Log).

27. Robert Dorr, "POWs in Iraq Survived Thanks to Training, Courage, Faith," *Naval Aviation News*, May–June 1991, 6.

28. History, 3 MAW, Marine Observation Squadron 1 (VMO-1) (Washington, DC: Marine Corps Historical Center, 1 January–28 February 1991).

29. Col Charles J. Quilter II, *US Marines in the Persian Gulf, 1990–1991: The Marine Expeditionary Force in Desert Shield and Desert Storm* (hereafter History, MEF) (Washington, DC: HQ USMC, History and Museums Division, 1993), 88; and History, 3 MAW, 166.

30. Gordon and Trainor, *Generals' War*, 369.

31. Ibid., 370.

32. Murray with Thompson, *Air War in the Persian Gulf*, 288.

33. History, MEF, 102.
34. Gordon and Trainor, *Generals' War*, 386.
35. Ibid., 385.
36. This was not logged on the USCENTAF/JRCC Incident/Mission Log.
37. MAJ Kevin Smith, *United States Army Aviation during Operations Desert Shield and Desert Storm* (Fort Rucker, AL: US Army Aviation Center, 1993), 93.
38. USCENTAF/JRCC Incident/Mission Log.
39. Message, JRCC A/R 251430Z, FEB 1991.
40. Gordon and Trainor, *Generals' War*, 387.
41. Edward Marolda and Robert Schneller Jr., *Shield and Sword* (Washington, DC: Naval Historical Center, Department of the Navy, 1998), 297.
42. Gordon and Trainor, *Generals' War*, 393.
43. Ibid., 396.
44. Murray with Thompson, *Air War in the Persian Gulf*, 292–94.
45. Capt Scott Fitzsimmons, interview by the author, 28 February 2002.
46. Lt Col Craig Mays, interview by the author, 22 March 2001.
47. USCENTAF/JRCC Incident/Mission Log.
48. History, 3 MAW, 166.
49. ARCENT G-3 Aviation After Action Report (Fort Rucker, AL: 5 April 1991).
50. Ibid.
51. USCENTAF/JRCC Incident/Mission Log.
52. Fitzsimmons, interview.
53. Ibid.
54. William F. Andrews, Oral History Interview (Colorado Springs, CO: US Air Force Academy, (USAFA), 25 September 1991).
55. Fitzsimmons, interview.
56. Ibid.
57. Capt Bill Andrews, USAFA oral history. Note: Capt Andrews was awarded the Air Force Cross for his actions on that day.
58. Fitzsimmons, interview.
59. Capt Gerry Stophel, interview by the author, 21 February 2002.
60. Capt John Steube, interview by the author, 9 January 2002.
61. Maj Lornie Palmgren, interview by the author, 2 May 2001.
62. Col George Gray, interview by the author, 3 May 2001.
63. Gray, interview; and e-mails by the author and Jim Blackwood on 25 February and 14 March 2002.
64. Lt Col Joe Hampton, interview by the author, 12 March 2000; and Steube, interview.
65. Gordon and Trainor, *Generals' War*, 412; and Maj Mason Carpenter, *Joint Operations in the Gulf War* (Maxwell AFB, AL: School of Advanced Airpower Studies, Air University, February 1995), 61.
66. Hampton, interview; and Durham, interview.
67. CW2 Steve Rogers, interview by the author, 2 May 2001.
68. James W. Bradin, *From Hot Air to Hellfire: The History of Army Attack Aviation* (Novato, CA: Presidio Press, 1994), 206.

69. Rogers, interview.
70. Bradin, *From Hot Air to Hellfire*, 208.
71. Ibid.
72. MAJ Mike Rusho, interview by the author, 10 April 2001.
73. Edward M. Flanagan Jr., *Lightning: The 101st in the Gulf War* (Washington, DC: Brassey's Press, 1993), 211.
74. Lt Col Richard Comer, interview by the author, 19 July 2000.
75. 1LT Al Flood, interview by the author, 15 January 2001; and e-mail from Al Flood to the author, 22 April 2002.
76. Ibid.
77. Ibid.
78. E-mail from LTC Bill Bryan to the author, 12 March 2002.
79. CPT Mike Thome, interview by the author, 14 July 2000.
80. MAJ Rhonda Cornum, interview by the author, 20 February 2001.
81. MAJ Rhonda Cornum with Peter Copeland, *She Went to War: The Rhonda Cornum Story* (Novato, CA: Presidio Press, 1992), 7.
82. Mark Hebein, interview by the author, 25 July 2001.
83. Jay A. Stout, *Hornets over Kuwait* (Annapolis, MD: Naval Institute Press, 1997), 219–23; and History, 3 MAW, 166.
84. Bradin, *From Hot Air to Hellfire*, 211.
85. Thome, interview.
86. Ibid.
87. Rogers, interview.
88. Stout, *Hornets over Kuwait*, 219–23; and History, 3 MAW, 166.
89. Hebein, interview.
90. Bryan, e-mail.
91. Steube, interview.
92. Message, JRCC A/R 281430Z.
93. USCENTAF/JRCC Incident/Mission Log.
94. Message, JRCC A/R 281430Z.
95. Marolda and Schneller, *Shield and Sword*, 305.
96. Gordon and Trainor, *Generals' War*, 410.
97. Perry D. Jamieson, *Lucrative Targets: The U.S. Air Force in the Kuwaiti Theater of Operations* (Washington, DC: AF History and Museums Program, US G.P.O., 2001), 160.
98. USCENTCOM/JRCC (Colorado Springs, CO: USAFA, Report for 28 February 1991).
99. Col Chuck Ware, USA, retired, phone conversation with the author, 21 January 2002.

Chapter 8

Postscript

The magic compass
—Nickname given to GPS receivers
by Arab ground commanders

Desert Storm was a short but violent conflict. In just 43 days, coalition forces destroyed Iraq's air force, a major portion of its army, and a large portion of the national infrastructure. Most importantly, they forced Saddam to pull his forces out of Kuwait. Our losses were relatively few, with 43 aircraft lost in combat.

Numerous potential loss incidents occurred throughout the conflict. JRCC personnel worked tirelessly to resolve every report and account for all aircraft and crews. For example, on 17 January, a signals intelligence agency reported an F-16 down. The JRCC controllers on duty called every F-16 unit in-theater to verify that all of their aircraft had safely returned from the day's missions. This was a routine and thankless task, but vitally necessary to account for all aircraft and personnel.

Given the total number of sorties flown, there were relatively few CSARs. What can be learned from this conflict? The record suggests the following:

1. The theater of operations was a challenging one for CSAR.

First of all, the area was mostly barren, affording little cover for evaders to exploit or for low-flying aircraft to terrain-mask. The terrain in the theater (Saudi Arabia, Iraq, and Kuwait) was a critical factor in this conflict. From the perspective of a shot-down airman, most of the terrain was flat or gently rolling, with almost no vegetation and nearly as hard as stone. This was in marked contrast to our experiences in Southeast Asia where so much of the area was covered in thick, triple-canopy jungle that afforded an evader excellent opportunity to hide if necessary.

This directly affected the survivors. As Col Ben Orrell pointed out, "Those guys over there—when they get shot down, they're standing out in the middle of a thousand miles of flat ground. There's no hiding."[1]

One British SAS trooper inserted into Iraq noted that, "The ground beneath us was dead flat and consisted of hard-baked clay, . . . and under that lay solid rock. There was no loose material with which to fill our sand bags."[2]

Both crew members of Slate 46, Lieutenants Jones and Slade, mentioned that they had a difficult time digging even small trenches in which they could recline. Several other survivors said that when they landed on the ground, there was just nowhere to hide.

There were a few exceptions. In northeast Iraq, there were some spectacular mountains, but no significant combat took place there. There were also some salt marshes along the Kuwaiti coast and one significant ridgeline just west of Kuwait City. And in the west, there were some valleys and wadis.

But for these exceptions, the terrain was generally level, sandy, and hard. This was great for interdiction, close air support, and the maneuver of mechanized forces, but not the best for downed airmen trying to hide.[3]

It was also a challenge to the helicopter rescue crews. Colonel Orrell explained, "In Vietnam, you had trees to protect you, hills and valleys. The stuff up there [Iraq] was just flat. They could see you coming from 100 miles away."[4]

The helicopter pilots talked of feeling naked as they flew over the vast expanses of the desert. Most felt compelled to fly as low as 10 feet to get any advantage at all from any irregularities in the terrain. That would afford them some protection from enemy threats, but also made it hard for them to electronically contact survivors.

The Iraqis took advantage of the openness. Knowing that we would try to rescue our aircrews, they sent fast-moving roving patrols over vast stretches to capture our men or ambush our helicopters.[5]

Secondly, the weather was harsh, and the local populace was generally supportive of Saddam. At this time of the year it was also extremely cold. Several British commandos were killed from exposure due to the cold temperatures, constant winds, and lack of cover. Additionally, the local people could not be trusted. Said one British officer, "You meet a Bedouin and you have got a fifty percent chance he is going to turn you in."[6] Maj Jeff Tice (Stroke 65) discovered this firsthand.

Additionally, Saddam published to his people a standing reward of $35,000 for the capture and presentation of any coalition airmen or soldiers to the Iraqi police or military. This was a fortune to most indigenous Iraqi people and strong incentive to most to comply.[7]

Thirdly, Iraqi air defenses were extensive and lethal. Iraq had one of the most sophisticated and intense air defense systems in the world. Composed of radars and communication equipment from several countries, it had almost 4,000 radar-directed SAMs, 7,500 heat-seeking SAMs, 972 fixed AAA sites, and over 8,500 AAA guns. Additionally, every tank and armored personnel carrier had at least one machine gun, and every troop had a rifle.[8] In toto, the skies of Iraq were a dangerous environment for coalition airmen.

2. This conflict reinforced the fundamental truth that the best CSAR device is to not get your aircraft shot down in the first place.

Given the realities of the theater noted above, this was a challenge, and some of the initial projections were that the coalition aerial forces would suffer horrendous losses. It did not happen. Several things helped to enable this.

Stealth Technology. The investments made in the 1980s in stealth technology paid off. No F-117s were shot down. They flew most of the missions into the most heavily defended areas. Brig Gen Buster Glosson noted, "As a result of our stealth technology, the Iraqis fired their AAA weapons at aircraft noise, not a visual or radar target."[9]

On 17 January F-117s attacked targets in Baghdad and suffered zero losses. More importantly, only one pilot per attacking aircraft was put at risk. Contrast this with a large package that attacked the Al Taqaddum Airfield that same day. Thirty-two F-16s dropped bombs supported by four EF-111s, eight F-4Gs, and 16 F-15s for MiG CAP. Seventy-two aircrewmen risked their lives in that one raid.[10]

Precision Weapons. The revolution in precision weapons was also a large factor. The ability to guide a single weapon to destroy a target instead of having to send numerous aircraft to try to hit it meant, quite simply, that fewer aircraft and aircrews were put at risk. The historical trends in weaponry are clear. In World War II, B-17s had a CEP (circular error probable—a statistical term meaning the radius from the target in which 50 percent of the bombs would fall) of 3,300 feet. To ensure the destruction of a specific target, 9,070 bombs had to be dropped. That required several hundred aircraft, each with a crew of 10 men. By the time of Desert Storm, an attack aircraft using laser-guided bombs had a CEP of 10 feet and could claim almost a 100 percent certainty of destroying the target while requiring just a few men.[11]

Better yet were the cruise missiles. Essentially unmanned aircraft, they could attack and destroy targets without putting any coalition aircrews at risk. Noted one campaign planner, "We used those weapons because . . . it seemed the logical thing to do. Plain and simple . . . it saved lives. If we had lost a half-dozen A-6s attacking those targets, it would have been unforgivable."[12]

Global Positioning System. The impact of the GPS on Desert Storm is just now being recognized. While not yet fully operational in 1991, the 16 satellites in orbit at the beginning of hostilities gave the coalition forces precision navigation capability for up to 22 hours per day. Unheralded are the efforts of USSPACECOM in general and the 2d Satellite Control Squadron in particular to keep the system functioning throughout the conflict, especially when three satellites suffered mechanical problems at various times and were in danger of losing their ability to transmit accurate signals.

Once the system was in place and a common reference datum had been applied, it gave our forces the ability to move throughout the theater quickly and accurately. It also provided our reconnaissance and intelligence forces—specifically the RC-135, AWACS, and E-8 Joint STARS aircraft—the ability to precisely determine the location of almost anything observable to 10-meter accuracy.[13]

The 72 Block 40 F-16C/Ds sent to the Gulf were noted for their bombing accuracy. GPS, combined with other highly specialized sensor pods on the aircraft, enabled them to accurately bomb day and night in all kinds of weather. All in all, GPS was one of the great enablers of our victory in Desert Storm and is labeled by many as "the unsung hero of the Gulf War."[14]

General Horner recognized this when he said, "During DESERT STORM, we used the Global Positioning System to solve the soldiers' most nagging battlefield problem—where am I? . . . While Iraqi forces were confined to charted roads and trails, Coalition forces had the run of the field. If you moved, or were in anything that moved in-theater, you wanted GPS with you."[15]

Electronic Warfare. The investments made in the area of electronic warfare—like the creation of the F-4G Wild Weasel fleet; the creation of the EF-111 fleet; electronic jamming pods like the ALQ-119, -131, -184, and -135; and the new antiradiation missiles—also paid off. It gave the coalition forces the ability to suppress the highly sophisticated Iraqi defenses to the point where aircraft could operate at medium levels with relative impunity.

Most aircraft were also equipped with the latest versions of radar-warning receivers (RWR). These devices could pick up enemy radar signals, thus warning the crews that they were being tracked by enemy radars. Such critical warning gave them a tremendous advantage in reacting to the threat.

Discussing the low losses, Horner said, "I would have to say it's one of the highlights of the war, especially if you look at the number of sorties we flew and the intensity of the air defenses. Our losses to surface-to-air missiles were something like ten planes." Horner went on to explain the value of area-jamming aircraft, "like the Air Force EF-111 and the Navy EA-6B, which pour electrons into the enemy's target-acquisition radars so he just doesn't know where you're coming from."[16]

But the airmen still had to deal with the guns. As one veteran of Desert Storm wrote, "USAF electronic superiority allowed air supremacy to be quickly achieved above 10,000 feet, but the numerous Iraqi AAA pieces and shoulder-fired surface to air missiles denied low-risk operations at lower altitudes."[17] There is just no way to calculate how many aircraft were not shot down because they could stay high.

Night Exploitation. Tremendous advances in night-vision technology and techniques allowed coalition forces to operate almost unrestricted at night. As one British pilot noted, "The first air strikes were to be carried out at night because of the superior night vision capabilities of the coalition aircraft. . . . It hid air activity on an enormous scale."[18]

POSTSCRIPT

Enemy gunners had a low probability of hitting what they could not see. This was another way of avoiding the AAA. Equipped with the latest generation of night-vision goggles, the Pave Low crews in particular were able to take advantage of this night capability.

Tactical Innovation. The airmen leading the coalition air effort showed great tactical flexibility. General Horner told his staff and the men in the TACC, "If you have a good idea about tactics or target selection or things of that nature, they are always welcome.... [There are] no bad ideas here.... Everybody has experience in one form or another in tactical aviation and we need to talk to one another about it."[19] For example, "Poobah's Party" was a brilliant operation that had a direct and decisive effect on the air campaign.

At several points the commanders were quick to adjust operations to increase effectiveness and avoid casualties. The restrictions on large-package attacks against Baghdad-area targets and the altitude restrictions were prime examples.

During the first three days of combat operations, one of the primary targets was the Iraqi air defense system. As that was being attacked, coalition losses were significant, with 17 fixed-wing aircraft downed, but the campaign was a success. One F-4G Wild Weasel pilot noted, "Our mission was accomplished in the first 48 hours of the war, especially that first night at Baghdad. We went in and destroyed so many of their surface to air missile/radar sites that they had a healthy respect for HARMs and Weasels. They were real hesitant to turn on their radars. After that, we basically just had to show up. They [Iraqi radar operators] would stay off the air when they knew we were present. They knew our call signs."[20]

In the last week, when the ground campaign was being conducted and all aircrews were instructed by the JFACC to "do what was necessary" to provide close air support, eight aircraft were lost.[21]

The British were also adept at changing their tactics. After the horrific losses suffered by their Tornado pilots on the dangerous but so necessary airfield attacks, they re-roled their aircraft and crews to operate in the medium-altitude regime. They lost only one more aircraft after that.

The AFSOC helicopter crews were also adaptive. They preferred, and intended, to do their rescue work at night. But when shoot downs occurred during the day, they quickly adapted to daylight work as part of a rescue task force, as seen with the Slate 46 recovery. To reduce response time, they would, in certain instances, launch to the border or hold in a safe area while the search process proceeded.

Training. In the 1970s, the USAF Tactical Air Command started a continuing series of "Flag" exercises. Red Flag was the best known. Conducted on the massive weapons ranges north of Las Vegas, Nevada, these exercises were initially designed to give each young fighter pilot his first 10 combat sorties in a forgiving environment where the price of failure was not death but learning.

One F-111 pilot said, "Training saved our lives. We trained for low and medium altitude war. . . . We fought like we trained."[22]

A Wild Weasel pilot noted, "I had trained eight years for this. . . . When the SA-2 launched, I didn't feel scared at all. I knew exactly what to do. The reason that we are all doing so well in this war is the fact that we are all so well trained."[23]

Red Flag helped the aircrews develop the critical situational awareness necessary to quickly and accurately understand what was going on around them in any combat situation. That dramatically reduced losses and saved lives.[24]

But the exercises did more that just train the young pilots. They trained commanders, too. They provided them with the opportunity to experiment with new tactics and techniques. They allowed them to mix combinations of aircraft to see what worked and what didn't. Red Flag was a giant laboratory for tactical tinkering, testing, and learning.

As Red Flag expanded, aircraft and crews from the Navy, Marines, and even the Army joined in, and the wealth of knowledge continued to expand. Eventually, even allies were included, and the experiments just grew.

There were others. Green Flag was just like Red Flag except the emphasis was on electronic warfare, specifically using Wild Weasels, EF-111s, and other electronic assets. Scenarios were designed so that they could practice and perfect their sophisticated electronic dances. It paid off over Iraq.

There was Maple Flag where various US, Canadian, and other air forces came together at Cold Lake, Alberta, Canada, and took advantage of the vast expanses to stage huge war games. In Blue Flag at Hurlburt Field, Florida, commanders and their command and control teams staged huge war games focused on ironing out the difficulties of trying to orchestrate such massive campaigns. The Internal Look exercise, where the Gulf War was prefought, was conducted at Blue Flag in the summer of 1990.

The author of the air campaign plan used in Desert Storm, Col John Warden, had predicted that overall losses would be about 40 aircraft. He stated that the Air Force (actually all joint and combined aerial forces) would use mass, shock, and the destruction of enemy command centers to gain air superiority and keep casualties low. His prediction was the most accurate.[25]

Overall, the low loss rate reinforced a trend reaching back to World War II, where the loss rate (expressed as number of aircraft lost per sortie) was 1.0 percent.[26] It dropped to .17 percent in the Korean conflict.[27] In the war in Southeast Asia, the overall loss rate was .081 percent.[28] The Gulf War rate fell to .055 percent.[29] All of the investments in stealth technology, precision weapons, and training had paid off.

3. Going into Desert Storm, Air Force CSAR capability had been dramatically reduced from the peak reached during the war in Southeast Asia.

Traditionally, the Air Force had been the primary supplier of air rescue capability, but between the Vietnam War and Desert Storm, this capability had been allowed to wane. This resulted from a series of force reductions, budget decisions, and reorganizations that culminated in the creation of the Special Operations Command (SOCOM), which absorbed some Air Force rescue elements. The Air Force leadership had taken steps to rectify the situation by initiating a rebuilding program in the newly established Air Rescue Service (ARS) just prior to Desert Shield/Storm. The ARS mission statement clearly indicated that it had CSAR responsibility worldwide, but it had only older assets of limited capability to perform the recovery part of the mission. The all-weather, night-capable helicopter that its predecessor had identified as critical to the mission after the war in Southeast Asia had been taken away and given to the AFSOC. New helicopters were being procured but were only available in limited numbers.

When the Klaxon rang for Desert Shield/Storm, the reconstituted ARS was able to provide some personnel for command and control (in the JRCC), but it had no force structure that could safely and reliably perform the combat recovery mission in high-threat areas. A provisional unit of HH-60s could have possibly been formed, but the ARS commander chose not to do so.

For this and other reasons, the mission passed to SOCCENT. Fortunately, there was a reservoir of individuals among the helicopter and MC-130 crews in the AFSOC community that had rescue experience. They were able to take the tasking and develop the capability for SOCCENT.

Navy combat recovery capability had been moved almost completely into its reserve component, but the two reserve units were well trained and filled by earnest men who would answer the call. When the Navy was granted authority by the president to recall reservists, activating and employing them made a great deal of sense and added to SOCCENT's combat rescue capability. They did not get any rescues, but did participate on numerous missions.

4. At the same time, the expectations among the flying crews for CSAR were very high.

This was based on the recorded history from Southeast Asia and war stories passed down by the veterans of that conflict to the young troops in the Gulf. These expectations were reinforced by specific comments of some senior Air Force officers.

The Vietnam generation of air warriors trained the Desert Storm crews. Around debriefing tables and bars from Korea to Germany, the "vets" inculcated the idea in the younger men that if they were shot down, the "Jolly Green" would be there. It was almost an article of faith.

But in the '80s, the specter of war across the plains of Europe caused a de-emphasis of CSAR. The crews realized that a war there would have been a war for survival, and the loss of whole squadrons could be expected. In Europe there were no rescue squadrons except for a unit in Iceland that was optimized for overwater rescue. The crews were taught that if downed, they were to evade to specific points. At specified times, SOF helicopters would fly by and pick up whoever was there.

Desert Storm was not a war for national survival. The crews saw that a combat rescue force was in place, and six CSAR exercises were conducted. So the old Vietnam stories were reinvigorated.

As combat loomed, Brig Gen Buster Glosson personally went out to speak to the aircrews. From his experiences in the last days in Southeast Asia, he felt an obligation to look them all in the eye and tell them why they were fighting. He told them that the nation was behind them. He told them that what they were doing was important and talked them through the air campaign plan. But then he cautioned them to be careful. This could last a while and we needed to husband our resources. He said, "There is not a . . . thing in Iraq worth you dying for."[30] He also told them that he would "stack helicopters on top of each other," to rescue them if they were shot down.[31]

This was an exact replay of the feelings among the aircrews in the last days of the Vietnam War. As one squadron commander there told his young pilots, "Guys, there is nothing over here worth an American life."[32]

Then the young captains from AFSOC went around and told them that if they got bagged, the rescue forces were coming—although with restrictions, and preferably at night. The history suggests that they also explained the necessary procedures.

One unit seemed to have gotten a different message. When Glosson visited the 4th Wing at Al Kharj, one young pilot remembered his saying that, "If you get shot down, you [won't] have to spend the night. If it's a night sortie, we'll be in the next day to get you, but you are not going to spend 24 hours on the ground."[33]

5. The non-rescue of an F-15E crew early in the conflict caused a morale problem among the F-15E units.

As the narrative of the Eberly and Griffith incident makes clear, the men of the 4th Wing were dramatically affected by the loss of Corvette 03 early in the war. This was a bad time for them. Just two nights prior, they had lost a crew (T-Bird 56) near Basra. Additionally, they were being whipsawed from target to target, often on the shortest of notice, as CENTAF tried to get a handle on the Scud missile dilemma. Such chaos was disconcerting to say the least. What is not clear is what specific CSAR briefings they were given before the war concerning location and authentication procedures or what classified intelligence they had available to them explaining the difficulty of getting rescue forces into that area of Corvette 03.

Colonel Gray at SOCCENT had what intelligence data was available and had to make the difficult decision on whether or not to send a helicopter in for them. The British commanders had to wrestle with the same issue when their team, Bravo 20, was compromised and split up.

But the men of the 4th Wing were not privy to any of that. Consequently, the rumors spread that the AFSOC guys would not go. Those feelings have made their way into the histories of the war and occasionally resurface in discussions about CSAR.

6. CENTAF did not have a quick, accurate, and reliable way to locate downed airmen or discretely communicate with them. There were several parts to this:

Obsolete Survival Radios. The PRC-90 survival radio most commonly issued to the aircrews—a holdover from the war in Southeast Asia—was far too limited in its capabilities and too easily exploited by the enemy. It was rugged and durable, but it was nothing more than a two-channel transmitter that could also transmit an emergency beacon signal. That signal was easily tracked by anyone who had direction-finding gear and knew the two internationally identified frequencies. It had no discrete capabilities either for communication or position identification. One planner at CENTCOM, LTC Pete Harvell, thought that the radio itself was a threat to the aircrews. He said, "The pilot becomes his own worst nightmare in that by trying to signal his position, he's jeopardizing himself and whatever rescue force could come to get him."[34]

Apparently, the RAF crews were carrying survival radios which were tuned to different frequencies than those used/monitored by US crews and rescue forces. Several RAF crewmembers mentioned that they made emergency calls or initiated their beacons with no response.

A new radio was available, the PRC-112, which had some discrete communication and location capabilities. The SOF community and the Navy had bought some and had them in the Gulf region in limited numbers for their troops. The Air Force had chosen not to purchase them before the war.

Additionally, to fully utilize the PRC-112, search aircraft had to be equipped with the DALS receivers to home in on the discrete signal it emitted. The Navy and AFSOC had installed DALS on their helicopters, but no service had installed it on any aircraft that could operate in a high-threat area. Two Navy fliers, LT Rob Wetzel (Quicksand 12 Alpha) and LT Larry Slade (Slate 46 Bravo), had PRC-112s when they went down. Wetzel left his behind when he tried to evade. Lieutenant Slade apparently had radio problems. That was particularly unfortunate because an MH-53 was in the area looking for him and had his discrete codes.

After the debacle with Corvette 03, Brig Gen Dale Stovall, the vice commander of AFSOC, personally called his counterpart at TAC and told him:

> I said, "Hey, we need to put one of these direction interrogators on an F-15E so we can get a fast mover out there and we can pick up where the guys are who get shot down. . . . The [HC-130] tankers can't be flying around at 10,000 to 15,000 feet out there, searching with their equipment for some guy on the ground and the helicopters can't pick up the signals when they are at 50 feet AGL." . . . I was told, personally told, "We don't have room to put it on the F-15E model. . . . You have to have fast movers doing the searching. . . . Once you find him, get him located, then you can launch the [rescue] force. . . . It doesn't do any good to have them [DALS] on the helicopter until you get into the terminal area."[35]

Perhaps a better plan would have been to use the Block 40 F-16C/Ds that had GPS. They were much more survivable in a high-threat area than any of the AFSOC aircraft. Modifying them with DALS receivers could have created a capability to accurately locate survivors with PRC-112s. This would also have been more effective for survivors with PRC-90s, since these aircraft could have marked the survivors' location with GPS. As noted, they were much more accurate than any INS used by most fighter aircraft. But the Block 40 F-16s were in high demand for attack missions. There is no indication that their use for CSAR search duties was ever considered.

AWACS, Rivet Joint, and Satellites. SARSAT satellites were available for survivor location, but their best accuracy was about 20 kilometers or 12 miles.[36] AWACS and Rivet Joint RC-135s also monitored for downed airmen, but their plotted positions were inaccurate, as seen in the Corvette 03 and Slate 46 shoot downs.[37] Their collection systems used GPS to establish their position. They were good at plotting enemy radio or radar transmitters, but accurately plotting the location of low-wattage PRC-90 and PRC-112 radios proved to be beyond their capabilities.

Limited Number of Survival Radios. Apparently, Air Force, Navy, and Marine aircrews were issued only one radio each. Uzi 11, Capt Dale Storr, had another radio in his personal survival kit, but it was tuned to the training frequencies. General Glosson did not make it a policy to have all crew members carry two radios. He said, "I did not know that my people were flying with one radio and it certainly was not my policy, I can assure you that. I left that solely up to the wing commanders and I intended to do that. . . . I just never thought that anybody would be short-sighted enough to fly with one radio in his vest. I just never thought that."[38]

Given the critical importance that the radio plays in the CSAR process, a repeat of the practice of issuing all aircrewmen two radios, as in the later years of the war in Southeast Asia, might have given some downed airmen at least a chance of being rescued in Desert Storm.

Failure to Exploit GPS for CSAR Purposes. Considering the impact of GPS on our overall operations in Desert Storm, the failure to extend that exploitation to CSAR was a missed opportunity. The overall GPS structure was in place. A common WGS datum had been applied. The primary combat recovery aircraft—the MH-53—had GPS navigational capability. But, as pointed out, accurate GPS navigation needs a known start point and *end point*. The recovery efforts for Corvette 03 and Slate 46 showed that the PRC-90 and PRC–112 radios could not be located that accurately with the assets in-theater and that coordinates reported by INS were not necessarily useable for GPS. The recovery of ODA 525 showed how beneficial it was for anyone being rescued to have the ability to report their GPS position. This was recognized by some; a few B-52 crews flying in Desert Storm had SLGRs packed in their ejection survival kits.[39] There is no indication or record that any tactical crews in Desert Storm were issued any model of GPS receiver.

Colonel Comer of the 20th SOS had perhaps the clearest vision of this. He recalls, "We didn't have the technology on our people who might be survivors to know their location. We had GPS. But GPS was brand new. We didn't understand that it changed the world: that it changed how you fly, that it changed where you fly. We did not know that yet."[40]

As one post–Desert Storm report said, "A glaring lesson learned in this conflict, and projected to be fixed by GPS technology, was the lack of precise position determination for downed pilots."[41]

The AFSOC helicopter pilots agreed with this overall assessment. As one later said, "This [issue] should constitute the number one lesson learned."[42]

7. Regardless, personnel from the service components and SOCCENT executed numerous joint missions and several SAR missions. Additionally, there were several intraservice rescues of note.

The JRCC log (table 1) shows 20 missions actually attempted from the beginning of hostilities until cessation on 28 February. But, as previously mentioned, not all rescue efforts were noted on the log.

Table 1. JRCC Log

Number	Call Sign	Recovery Element	Outcome
JM001	Jupiter	USMC	None, safe landing
JM002	Hostage 75	USN/RSAF/RSN	Search, nonrecovery
JM003	Stroke 65	SOC	CSAR, nonrecovery
JM004	Dark 15	Saudi	SAR, recovery
JM005	Stamford 11	Saudi	SAR, recovery
JM006	Slate 46	SOC	CSAR, one recovery
JM007	Corvette 03	SOC	CSAR, nonrecovery
JM008	Wolf 01	USN	CSAR, recovery
JM009	Cat 36	USMC	Search, nonrecovery
JM010	(Jaguar)	RAF	SAR, non-event
JM011	Spirit 03	SOC/USN/Saudi	Search, nonrecovery
JM012	Uzi 11	SOC	Nonrecovery
JM013	Millcreek 701	USMC	SAR, nonrecovery
JM014	Jump 57	SOC	Search, nonrecovery
JM015	Hunter 26	SOC/Saudi	Search, nonrecovery

Table 1. (continued)

Number	Call Sign	Recovery Element	Outcome
JM016	Enfield 37/38	SOC	Search, nonrecovery
JM017	Benji 53	SOC	CSAR recovery
JM018	Pride 16	AF/USMC	Search, nonrecovery
	Jump 42	USMC	CSAR, ground team recovery
	(AH-64)	Army	CSAR, ground team recovery
JM019	Magic 14	SOC	Search, nonrecovery
JM020	Mutt 41	Army	CSAR, nonrecovery
	Bengal 15	Army	CSAR, nonrecovery

Overall, coalition combat losses were 43 aircraft, recorded as follows:

Fixed-wing aircraft[43]		Rotary-wing aircraft[44]	
USAF	14	US Army:	
USN	6	AH-64	1
USMC	7	OH-58D	1
RSAF	2	UH-1	1
RAF	7	UH-60	2
Italy	1	TOTAL	5
Kuwait	1		
TOTAL	38		

Table 2 analyzes those combat losses for which *any* amount of significant data is available. It shows that, in fact, 87 coalition aircrewmen were downed in combat. Of that total, 48 were killed (55.1 percent), one remains classified as MIA (1.1 percent), and 38 survived the shoot down (44.1 percent). Of the 38 who survived the shoot down, eight were rescued (9.3 percent of total) and 30 were POWs (34.4 percent of total).

Table 2. Analysis of combat losses

Date	Aircraft/Service	Call Sign/Crew Size	Result/Target/Rescueable
17 Jan	F-18 /USN	Sunliner 403/1	1 KIA (MIA*)/escort
17 Jan	Tornado/RAF	Norwich 02/2	2 POW/airfield attack/no
17 Jan	A-4/Kuwait AF	Bergan 23/1	1 Evade-POW/enemy forces/no
17 Jan	Tornado/RAF	Norwich 21/2	2 KIA/airfield attack

Table 2. (continued)

Date	Aircraft/Service	Call Sign/Crew Size	Result/Target/Rescueable
17 Jan	F-15E/USAF	T-Bird 56/2	2 KIA/airfield attack
17 Jan	A-6/USN	Quicksand 12/2	2 POW/airfield/possible
18 Jan	Tornado/Ital. AF	Caesar 44/2	2 POW/airfield attack /no
18 Jan	OV-10/USMC	Hostage 75/2	2 POW/enemy troops/no
18 Jan	A-6/USN	Jackal 11/2	2 KIA/enemy port
19 Jan	F-16/USAF	Stroke 65/1	1 POW/Baghdad/no
19 Jan	F-16/USAF	Clap 74/1	1 POW/Baghdad/no
19 Jan	Tornado/RAF	Newport 15B/2	2 POW/airfield attack/yes
19 Jan	F-15E/USAF	Corvette 03/2	2 POW/Scud sites/possible
20 Jan	F-14/USN	Slate 46/2	1 POW, 1 Save/ escort/possible
22 Jan	Tornado/RAF	Stamford 01/2	2 KIA/radar site
23 Jan	F-16/USAF	Wolf 01/1	1 Save (fratricide)
24 Jan	Tornado/RAF	Dover 02/2	2 POW/airfield attack/no (fratricide)
24 Jan	F-18/USN	Active 304/1	1 Save/interdiction
28 Jan	AV-8/USMC	Cat 36/1	1 POW/enemy troops/no
31 Jan	AC-130/USAF	Spirit 03/14	14 KIA/enemy forces/no
02 Feb	A-6/USN	Heartless 21/2	2 KIA/crashed at sea
02 Feb	A-10/USAF	Uzi 11/1	1 POW/enemy forces/no
05 Feb	F-18/USN	WarParty 01/1	1 KIA/crashed at sea
09 Feb	AV-8/USMC	Jump 57/1	1 POW/enemy forces/no
13 Feb	F-5/RSAF	Hunter 26/1	1 POW/enemy forces/possible
13 Feb	EF-111/USAF	Ratchet 75/2	2 KIA/support/no
14 Feb	Tornado/RAF	Belfast 41/2	1 POW, 1 KIA/airfield/no
14 Feb	2 A-10s/USAF	Enfield 37 & 38/2	1 POW, 1 KIA/enemy forces/no
16 Feb	F-16/USAF	Benji 53/1	1 Save
19 Feb	OA-10/USAF	Nail 53/1	1 POW/enemy forces/no
20 Feb	OH-58/US Army	Tango 15/2	2 KIA/recce/no
23 Feb	AV-8/USMC	Pride 16/1	1 KIA/enemy forces/no
25 Feb	OV-10/USMC	Pepper 77/2	1 POW, 1 KIA/enemy forces/no
25 Feb	AV-8/USMC	Jump 42/1	1 Save/enemy forces
25 Feb	AH-64/US Army	Unknown/2	2 Saves/enemy forces

Table 2. (continued)

Date	Aircraft/Service	Call Sign/Crew Size	Result/Target/Rescueable
27 Feb	UH-1/US Army	Unknown /4	3 KIA, 1 Save/support
27 Feb	UH-60/US Army	Unknown /6	6 KIA
27 Feb	AV-8/USMC	Magic 14/1	1 KIA/enemy forces
27 Feb	F-16/USAF	Mutt 41/1	1 POW/Rep Guard/no
27 Feb	UH-60/US Army	Bengal 15/8	3 POW, 5 KIA/Rep Guard/no

*In 2001, Pres. Bill Clinton changed the status of the pilot, LCDR Michael Speicher, to "Missing In Action." See http://www.nationalalliance.org. In 2002, the secretary of the Navy changed Speicher's status to "Missing/Captured." See http://www.nationalalliance.org/gulf/secnavmemo.htm.

Among the POWs, who was rescueable? The official SOCOM history of the war states: "How many guys were 'rescueable'? Many [aircrewmen] landed in areas of heavy Iraqi troop concentrations or even airfields and the enemy troops were able to grab them before any effort could be made in their behalf."[45] Yet, one of the RAF pilots, Flt Lt Rupert Clark, stated that 30 minutes after he was shot down, he was in the base commander's office.[46]

The A-10 pilots and AV-8 pilots were almost all shot down right over enemy field units and captured within moments of landing. One AV-8 pilot remembered that, "Based on our targets, we knew that we were going to be right in the midst of a bunch of . . . people."[47]

There is no military definition of what "rescueable" means. Perhaps an objective definition would be that for a period of time, they were down in an enemy-controlled area but not under enemy control. It is hard to determine in each case what that period was, but there were a few cases where survivors—for some period of time—were in this situation.

Using that criterion, the recovered data shows that possibly eight of the 30 crew members who became POWs were rescueable. That is 9.2 percent of the 87 crew members shot down. Each situation deserves a review:

Hunter 26 (RSAF F-5). The pilot was down just north of the Iraqi border. Rescue forces were in the area but did not make contact for an unrecorded reason. He was reportedly captured by the Iraqis.

POSTSCRIPT

Source: 16 SOW History Office, Hurlburt Field, FL

Capt Tom Trask, who led one mission to rescue the crew of Slate 46, is shown here as a lieutenant colonel and commander of the 20th Special Operations Squadron. He is one of several squadron leaders who gained valuable experience flying CSAR in Desert Storm.

Quicksand 12 (A-6 with a crew of two). The loss of their two radios took away their ability to communicate with the rescue forces before they were captured within an hour. They were down in a dangerous high-threat area. Any attempt to do any type of visual search would have been extremely risky.

Newport 15B (RAF Tornado with a crew of two). The crew was in enemy territory for an estimated 14 hours, much of it at night, before they were captured. Repeated calls on their survival radios and emergency beacons were not detected.

Corvette 03 (F-15E with a crew of two). The inability to accurately locate the crew, coupled with the political problems of the overflight of Syria, delayed the rescue forces. Once the political hurdles were cleared, the rescue forces tried valiantly to get them. A Pentagon review of the Corvette 03 mission noted:

> Complicating the entire effort were unreliable intelligence inputs. There was never an accurate position fix and only one radio call was ever

received authenticating 03's status (on night two). This is the result of no reliable, secure, radio/position locating device issued to aircrew or compatible aircraft mounted locating equipment with any range and accuracy.

Syrian intransigence on over-flight delayed any effort at least 2 days. Had they been amenable, search aircraft would have at least been launched to Syrian airspace in night two to fix 03's position and initiate a possible rescue.[48]

Slate 46 (F-14 with a crew of two). The pilot was rescued. The RIO (radar intercept officer), Lieutenant Slade, could not contact the rescue forces with his radio. MH-53s from the 20th SOS entered the area twice but were unable to contact him.

In a best-case computation, if all eight above were truly rescueable, then adding in the number who actually were recovered totals 16 crew members. That means eight out of 16, or 50 percent, of those who were rescueable actually were rescued.

Of the eight who were rescued:

- Two were recovered by helicopters on land (Slate 46 Alpha and Benji 53)
- One was rescued by special operations forces at sea (Wolf 01)
- One was rescued by the Navy at sea (Active 304)
- One was recovered by a USMC ground team (Jump 42)
- One was recovered from a UH-1 by a US Army ground team
- Two were recovered from an AH-64 by another helicopter

8. At no time were SOF aircraft "not available" for rescue missions.

The initial mission assigned to SOCCENT was combat recovery. Immediately upon arriving in-theater, MH-53s were put on alert. Additionally, the first mission given to the Army Special Forces teams and the Navy SEAL teams was combat recovery.

The AFSOCCENT, Col George Gray, and his director of operations, Col Ben Orrell, both stated emphatically that their helicopters were on alert at several locations throughout the war. At no time did other SOF tasking interfere with this alert status. That included helicopters from the Navy (HCS-4/5) and the Army (the 3d/160th).[49] The 3d/160th had come over expecting special operations missions and had, at one point, complained

POSTSCRIPT

about being held on combat rescue alert at the expense of other tasking. The rescue of Benji 53 changed their thinking.[50]

Specifically, Colonel Orrell said, "There were times when we had a certain number of aircraft committed to [rescue] and Colonel Johnson would be planning a Special Ops mission. We'd have to pare down the number of aircraft we would have used based on the fact that we had to have them on standby for CSAR."[51]

When asked if combat rescue was a secondary mission for JTF Proven Force, Capt Steve Otto, from the 21st SOS at Batman, stated, "That is absolutely false. For my firsthand experience with the 21st, we never came off of CSAR alert. We never had any special operations forces tasking. [Combat rescue] was our mission."[52]

As for SOCCENT's dedication to the combat recovery mission, Col George Gray stated after the war, "It was understood. That was my thing. I was the Czar. I was the guy. We set everything up as best as we thought we could do it. . . . When we deployed out of Hurlburt for Desert Storm, our first and foremost mission was [combat rescue]. We developed a briefing and took it to Horner. He blessed it. We took a crew around and briefed every single unit in-theater that was going to be flying a combat mission up north."[53]

On some occasions, though, SOF aircraft were not dispatched by the SOF commanders because they assessed the mission as just too dangerous. Although coalition forces had quickly attained air superiority in-theater, this had different meaning for fixed-wing aircraft that can operate above 10,000 feet and helicopters that, by their nature, have to operate at low levels. Noted one helicopter pilot, "Anytime you're below 1,500 feet or above three feet, you're in what we call the 'dead man's' zone."[54]

Lt Gen Fred McCorkle, the deputy commandant of the Marine Corps for aviation, said, "Unlike jet bombers and fighters, choppers often have to hover or land in hostile territory to do their jobs. It is a pretty dangerous place to be, and you're vulnerable. When you can fly 30,000 feet over the enemy, you can feel pretty confident that you aren't going to be shot. That isn't true when you're flying at 1,000 feet or coming into an LZ [landing zone]. You never know who's going to be down there that can do you in."[55]

Col Bennie Orrell understood the problem. He said:

> I'm not kidding you; you could see those Paves [MH-53s] 50 miles off. There was no hiding them. That's a big ol' slow moving target—I was reluctant to go cruising in there in the daytime. There certainly may have been a situation where we would have done it, but if you don't have a guy talking to you on the radio, it's pretty hard to convince me to send another two or three crews in there. . . . The only way we were going to survive as a rescue force in that environment was to fly at night. And I don't think that the fast movers [fighter pilots] ever accepted the fact that we were not just going to come plunging in there in daytime like we had done in Vietnam. Had we done that, we'd have lost more than just that one [Bengal 15]. We would have lost lots of them. It's a tough call to make—whether George [Gray] is making it or whomever is making it—to say that we can't go until dark. . . . I salute George for making those calls because there simply was no way to survive in that environment.[56]

But helicopters were dispatched into Iraq for search purposes when the threat assessment allowed, as was done for the Bravo 20 team. All in all, Bennie Orrell felt that George Gray was calling the missions just about right.[57]

It should be noted that during the conflict, *zero personnel were lost or killed on rescue missions directed by SOCCENT*. Bengal 15, a line Army crew, was shot down on a rescue mission, but they were dispatched from the TACC.

Recalling the BioTechnology study cited in chapter 1, Air Force and Navy rescue forces rescued 778 downed aircrewmen throughout Southeast Asia. In the process, 109 aircraft of all types were shot down and 76 rescue personnel were killed or captured. That meant that one rescue or support aircraft was lost for every seven men saved, and one rescue troop was lost for every 10 men saved. That was a high price to pay for rescue.

Col Jesse Johnson and Col George Gray personally oversaw any commitment of SOCCENT assets to combat recovery missions. Most, they approved; some were turned down. Did they call it right? Does this mean that they were far too quick to withhold committing helicopters and crews?

As opposed to Southeast Asia, where the losses of rescue personnel were significant, SOCCENT had a zero loss rate. This met President Bush's objective of minimum casualties.

Bengal 15 was shot down on 27 February when an Army crew tried to rescue USAF Capt Bill Andrews. What is instructive about that mission is that when the SOCCENT commanders were called, they analyzed the mission and turned it down, realizing it was far too dangerous for a daytime rescue. The rescue system was then short-circuited, and a helicopter—not from a combat-recovery-trained and dedicated unit, but from an Army line battalion—was launched from within the TACC into an area that several people had said was too dangerous. The result was predictable. What George Gray and Jesse Johnson knew would happen did, in fact, happen. This seems to indicate that they were calling the missions about right.[58]

It all had to do with air superiority, of which General Horner said:

> [A]ir superiority is not a precise concept. And the process of gaining it is no less fuzzy. What do you mean by air superiority, and how do you know when you have got it? There is no handy chart that lets you plot the x- and y-axis and find where the two lines cross. What I wanted to do was operate freely over Iraq and not lose too many aircraft. Okay, what does that mean? What is too many? . . . Free operation over Iraq raises other issues. For starters, not every aircraft could be expected to go everywhere. Or if it could go everywhere, it might not do that all the time. . . . In other words, control of the air is a complex issue, filled with variables.[59]

To the fighter guys and anyone else who could operate above 10,000 feet, the suppression of enemy MiGs and radar-guided SAMs allowed them to do their jobs with relative impunity. But the helicopter guys had to operate down low. They weren't nearly as worried about MiGs and SAMs as they were about enemy guns and rifles or heat-seeking SAMs. That was their threat. Colonel Gray had tried to make this point to the planners during Desert Shield.

As one rescueman wrote, "If you run across a platoon size of deployed guys, they can shoot your butt right out of the sky, and God only knows we couldn't track divisions much less platoons. A squad could put maybe a couple of rounds in you. . . . a platoon, 35 or 40 guys, would with AK-47s just eat you up."[60]

The AFSOC guys seemed to have a better understanding of the elements of risk and reward in the combat recovery mission. One of them wrote, "If I thought the crews were doing anything unsafe I would stop them. . . . Be . . . sure you recognize the difference between taking an acceptable risk and an outright unsafe situation. That is where common sense and judgment are critical and experience must be your guide."[61]

One of the downed airmen himself, LT Jeff Zaun, understood the problem clearly. He said of rescue, "the place that we were—I could not get there at 400 knots. So how is a CSAR [helicopter] going to get in at 50 or 100 knots? . . . You are telling me that somebody is going to come in in a helicopter?"[62]

It was expressed even better by a fighter pilot who wrote after the war that, "USAF electronic superiority allowed air supremacy to be quickly achieved above 10,000 feet, but the numerous Iraqi AAA pieces and shoulder-launched [heat-seeking] surface to air missiles denied low-risk operations at lower altitudes."[63]

The man who wrote that was Maj Bill Andrews in his thesis at the School of Advanced Airpower Studies (SAAS) at Maxwell AFB in 1995. He had license to write it; he had flown as Mutt 41 on 27 February 1991. It was his F-16 that was blown out of the sky by a heat-seeking SAM. It was his rescue helicopter that was destroyed by Iraqi AAA.

The history of CSAR in this conflict seems to suggest that there are gradations to air superiority—that air superiority for an F-15 is far different than air superiority for an MC-130, MH-53, or UH-60. This history also seems to suggest that the AFSOC guys had the best comprehension of what the rescue forces (helicopters) could and could not do, and they exercised control over them to prevent what they felt would be unnecessary losses.

But Brig Gen Buster Glosson felt differently:

> If I'm willing to lose [helicopters] as the commander, I should have the prerogative to send that helicopter in there or send two or three in there, understanding I may lose one of them. That's my decision. It should not be someone else's decision. Who . . . anointed AFSOC and promised them that they would have no losses in war? That the only losses were going to come from the bomber pilots and the fighter pilots? Where . . . did this philosophy come from? I am not saying you send people into harm's way just to say you did it. But many times . . . you can assist the CSAR effort with distractions in a way that a helicopter can sneak in and not have near the exposure.[64]

9. Making CENTAF responsible for CSAR and then withholding operational control of the combat recovery assets violated the principle of unity of command.

Under USCINCCENT OPLAN 1002-90, General Horner, as COMUSCENTAF, was the designated SAR coordinator (SC) with overall responsibility and authority for operation of the JRCC and for joint SAR operations within the assigned geographical area. As such, he was responsible for CSAR. His JRCC, located within the TACC, had the authority to call upon any asset that they controlled to do what needed to be done to rescue coalition personnel down in enemy territory. This specifically included all of the reconnaissance and intelligence assets needed to locate the survivors, but they did not have operational control of the recovery helicopters. That operational control resided with SOCCENT and the other service components.

This greatly concerned Buster Glosson. He recalled, "You may say to Horner, 'you have the responsibility for CSAR and SAR,' but let there be no doubt, he was not confused about being totally in charge when he had to call somebody and say, 'Would you please do the following?' You are in charge when you call somebody and say, 'I want three helicopters on the island in the center of the lake, 27 miles on the 236° radial from the center of Baghdad, and I want them there tonight at 11 o'clock.'"[65]

The operational orders and supplements confused the issue, for they seemed to imply that SOCCENT was responsible for CSAR overall. This concerned Col George Gray. Several times, he pointed out that his helicopters were combat rescue assets and that SOCCENT could not accomplish any of the other functions necessary to prosecute CSAR. All of that needed to be accomplished by the JRCC working with the TACC. He and the SOCCENT commander, Colonel Johnson, established three criteria for launching their helicopters:

1. A location of the survivor(s).
2. Evidence of aircrew survival:
    a. Visual parachute sighting, and/or
    b. Voice transmission from the crewmember and authentication.
3. A favorable enemy threat analysis.

Practically speaking, this meant that the JRCC did not have control of all the assets it needed to recover personnel in enemy territory, because SOCCENT would not launch helicopters until it was satisfied that the mission was doable and surviveable for its crews. This caused ill feelings between the JRCC and

SOCCENT. The JRCC guys, Colonel Hampton in particular, wanted to launch helicopters immediately. He said that, "During Desert Storm, I would have liked to have the authority to launch recovery assets immediately. But we did not own the assets, so you couldn't just say 'go launch.' . . . Our concern was to get the guy, to get there as soon as possible."[66]

At several points during the Corvette 03 and Mutt 41 events, JRCC personnel felt that the SOCCENT guys were not being responsive or, worse yet, saw the JRCC, not as an overall theaterwide headquarters, but just an RCC for the Air Force.

The SOCCENT guys resented that point of view. They reciprocated the ill feelings. One SOCCENT commander noted, "We didn't like JRCC because they didn't seem to know what . . . they were doing. We did not understand what was going on in their minds and they did not understand us. They didn't trust our judgment."[67]

Yet it must be noted that none of the JRCC personnel had any CSAR experience or had received any significant CSAR training prior to deployment.[68] To Lt Col Rich Comer, this was one of the main deficiencies in our CSAR program in Desert Storm.[69]

Supporting this, Maj John Steube, the USMC augmentee to the JRCC, said, "In the JRCC, nobody had any combat experience. There wasn't a single person there who was a combat SAR type. They were all the Scott AFB guys and they had two Navy liaisons and two Marine liaisons and none of us had any combat SAR experience. . . . There were a lot of shortcomings there. It was obvious to me that CSAR was an afterthought in the whole DESERT STORM experience."[70]

Additionally, the JRCC commander, Lt Col Joe Hampton, was never able to get the necessary communications links into the JRCC or clearances necessary so that they could directly access more highly classified information.[71]

Whether or not General Horner and hence the JRCC should have had operational control of the helicopters for CSAR is a valid question. The JRCC was also the Air Force RCC. If the Air Force had deployed helicopters to the Gulf that were combat-recovery capable, they could have been directly tasked by them, and this issue would not have arisen. However, in the 1980s the Air Force moved the aircraft most capable of doing that mission to the AFSOC. That command now belonged to the

SOCOM, and its units were deployed to the Gulf under SOCCENT. The ARS had less-capable aircraft and had just received new aircraft, which could have been deployed in provisional units. Its commander chose not to do so.

So the question hangs: If the JRCC had direct tasking authority over the SOCCENT helicopters, would that have made a difference for the crew of Corvette 03, or Quicksand 11, or Newport 15 Bravo, or Slate 46 Bravo? Would this change have improved the capability of the JRCC to locate its survivors?

To one helicopter pilot, LT Rick Scudder on the USS *Saratoga*, having the special operations forces guys do the combat recoveries made a lot of sense. He remembered that:

> It's everybody's job but there is a great gulf between normal operating forces which was—then and now—the conventional forces, AF rescue, Army to a certain extent—they teach personnel recovery. It's not really a hard core mission of [the Army's], but they have tremendous capabilities... and then there is [sic] the capabilities of SOF. They train to it. They do it. They have aerial refueling capabilities. They have integration capabilities with other types of special operations forces that are a total breed apart in terms of philosophy, in terms of the threat they can legitimately and prudently counter, in terms of the risk they are willing to mitigate because of their extra training, their specialization, their special equipment, the tactics and techniques that they have and use. I have no bones with that.[72]

The fact remains that during Desert Storm, there was not one headquarters responsible for and in control of all the assets necessary for CSAR—location, authentication, and recovery. To Buster Glosson, this was a travesty. He commented,

> I think that not having the CSAR under the direct tasking order of the air component commander is the dumbest thing I have ever seen. Give me a break! We chop [assign] all air related weapons systems in a theater to the air component commander except CSAR? Now how stupid is that?... If those assets are going to be used for other than CSAR purposes, they should be under the special ops commander. But when there is a CSAR requirement whether they go or not should be at the discretion of that air component commander and not at the discretion of some other commander. It is just not right. It defies all logic of chain of command and jointness.[73]

In retrospect, General Glosson offered a sober summation. Regarding the SOCCENT helicopter crews, he said, "I want to make this abundantly clear: there has never been a situation where there were more forward-leaning, tougher and braver people than the guys flying the CSAR [missions] in Desert Storm.

I have nothing but the highest, total respect for them and their willingness to do anything that I asked."[74]

Addressing the command and control problem, he continued, "The problem is at the commander level or the number two guy that wanted assurance before they would generate much effort. You can't do that in CSAR if you are going to expect to be successful. You will minimize the losses of the people trying to do the CSAR but you will lose a lot of people you should have picked up. And that is exactly what we did in Desert Storm."[75]

So who was right, General Glosson or Colonel Gray? The loss of Bengal 15 seems to support Gray. He and his senior airmen were the experts on combat rescue. Within their community resided vast amounts of "hands-on" experience hard learned in the killing zones of Southeast Asia. Guys like Dale Stovall, Benny Orrell, and Rich Comer had been there. They had been in the hover while the bullets ripped. They knew what helicopters flown by the bravest of crews could and could not do. Yet General Glosson had a theater air campaign to wage and offered another perspective: "You can't always be so precise and so regimented that you never have confusion or, as they say, 'fog of war.' If you do, you are going to become so regimented that you are predictable. Commanders must keep the correct balance."[76]

Placing all of this historically, Glosson offered, "CSAR was broke before the war because of a mismanagement and a failure of the Air Force leadership. It was broke during the war because of the CINC's acquiescence to not use his fist and an anvil to correct it, in the interest of harmony."[77]

His commander, General Horner, seems to share his views. But even though Horner was himself critical of the overall CSAR effort in the war, he offered what appears to be the best single assessment of CSAR in Desert Storm when he wrote, "In fairness to the special operations forces [SOCCENT] commanders, the paucity of CSAR missions can't be blamed entirely on them. First, the density of Iraqi air defenses has to be taken into account. Flying a helicopter into a near-certain shoot-down obviously made no sense. Second, several pilots were captured shortly after parachuting over the Iraqi Army units they'd just attacked. Third—and most important—few aircraft were actually lost in combat. Thus, little CSAR was actually needed."[78]

In addressing the lack of unity of command, he seemed to support the feelings of Buster Glosson when he added, "The next Chuck Horner to fight an air war had better pay close attention to the way he (or she) organizes and controls the employment of his or her combat search and rescue efforts."[79]

Overall in Desert Storm, CSAR appears to have been a mixed bag. Because of advances in precision weaponry, satellite-based GPS technology, countermeasures, and training, relatively few coalition aircraft were shot down. Forty-three coalition aircraft were lost in combat, mostly over high-threat areas. Eighty-seven coalition airmen, soldiers, sailors, and marines were isolated in enemy or neutral territory. Of that total, 48 were killed, one is still listed as missing, 24 were immediately captured, and 16 were exposed in enemy territory. Of those who survived, most landed in areas controlled by enemy troops. Of the few actually rescueable, eight were not rescued for a host of reasons, but primarily because of limitations in CENTAF's ability to locate them accurately and in a timely manner. Additionally, a lack of unity of command over all CSAR elements caused confusion and a strained working relationship between the JRCC and SOCCENT, which operationally controlled the actual combat rescue assets. Regardless, the men who actually carried out the rescue missions displayed a bravery and élan so common to American rescue forces in earlier wars.

After Desert Storm, these issues were examined in detail and many changes were or were not made. In the not-too-distant future, though, AFSOC forces would be used again to perform the combat recovery mission in conflicts in the Balkans. Like a phoenix, many of these same issues would resurface. But that is another story to be told by another author.

Notes

1. Col Ben Orrell, interview by the author, 17 January 2002.
2. Chris Ryan, *The One That Got Away* (London, UK: Century Press, 1995), 41.
3. Perry D. Jamieson, *Lucrative Targets: The U.S. Air Force in the Kuwaiti Theater of Operations* (Washington, DC: AF History and Museums Program, Government Printing Office, 2001), 8.
4. Orrell, interview.

5. GEN H. Norman Schwarzkopf with Peter Petre, *It Doesn't Take a Hero* (New York: Bantam Books, 1992), 418.

6. Tom Clancy with Gen Carl Stiner, *Shadow Warriors* (New York: G. P. Putnam's Sons, 2002), 437.

7. Amy W. Yarsinske, *No One Left Behind: The Lt. Comdr. Michael Scott Speicher Story* (New York: Dutton Books, 2002), 106.

8. "Desert Storm, Final Report to Congress: Conduct of the Persian Gulf War 1991," *Military History Magazine*, April 1992, 177.

9. Rick Atkinson, *Crusade: The Untold Story of the Persian Gulf War* (Boston, MA: Houghton Mifflin, 1993), 259.

10. Williamson Murray with Wayne Thompson, *Air War in the Persian Gulf* (Baltimore, MD: Nautical and Aviation Publishing Company of America, 1996), 128.

11. Michael R. Gordon and Gen Bernard E. Trainor, *The Generals' War: The Inside Story of the Conflict in the Gulf* (New York: Little, Brown and Co., 1994), 189.

12. John Tirpak, "The Secret Squirrels," *Air Force Magazine*, April 1994.

13. Michael R. Rip and James M. Hasik, *The Precision Revolution: GPS and the Future of Aerial Warfare* (Annapolis, MD: Naval Institute Press, 2002), 188.

14. Ibid., 121.

15. Ibid., 117.

16. James W. Canan, "The Electronic Storm," *Air Force Magazine*, June 1991.

17. Lt Col William F. Andrews, *Airpower against an Army* (Maxwell AFB, AL: Air University Press, 1998), 35.

18. Charles Allen, *Thunder and Lightning: The RAF in the Gulf: Personal Experiences of War* (London, UK: Her Majesty's Stationery Office, 1991), 52.

19. Andrews, *Airpower against an Army*, 57.

20. Tim Ripley, "Desert Weasels," *United States Air Force Yearbook 1992*, 60.

21. "Desert Storm, Final Report," 179.

22. Murray with Thompson, *Air War in the Persian Gulf*, 79.

23. Ibid.

24. Tom Clancy with Gen Chuck Horner, *Every Man a Tiger* (New York: G. P. Putnam's Sons, 1999), 355.

25. Gordon and Trainor, *Generals' War*, 90.

26. Clancy with Horner, *Every Man a Tiger*, 502.

27. Ibid.

28. Ibid.

29. Thomas A. Keaney and Eliot A. Cohen, *Gulf War Air Power Survey* (hereafter *GWAPS*), *Summary Report*, vol. 5, pt. 2 (Washington, DC: Government Printing Office, 1993), 116.

30. Al Santoli, *Leading the Way, How Vietnam Veterans Rebuilt the U.S. Military: An Oral History* (New York: Ballantine Books, 1993), 205.

31. Brig Gen Buster C. Glosson, interview by the author, 25 September 2002.

32. Darrel D. Whitcomb, *The Rescue of Bat 21* (Annapolis, MD: Naval Institute Press, 1998), 142.

33. 1st Lt Jeff Mase, interview by the author, 9 April 2001.
34. LTC Pete Harvell, interview by the author, 29 January 2002.
35. Brig Gen Dale Stovall, interview by the author, 3 September 2001.
36. ICAO Circular 185-AN/121, *Satellite Aided Search and Rescue–The COPAS SARSAT System* (Montreal, Canada: International Civil Aeronautical Organization, 1986), 17.
37. Lt Col Richard Comer, interview by the author, 19 July 2000; and Gordon and Trainor, *The Generals' War*, 263.
38. Glosson, interview.
39. Rip and Hasik, *Precision Revolution*, 146.
40. Comer, interview.
41. CDR Patrick Sharrett et al., "GPS Performance: An Initial Assessment," *Navigation Journal*, vol. 39, no. 1 (Spring 1992), 405.
42. Comments by Maj Gen Rich Comer on this book, 2 February 2003.
43. *GWAPS*, 114.
44. Mr. James Williams, Info paper (ATZQ-AP-G), Subject: Aviation Units deployed (Fort Rucker, AL: US Army Aviation Center, 17 April 1991).
45. Capt William LeMenager, HQ SOCOM History, "A Gulf War Chronicle," unpublished manuscript, January 1998, 37.
46. Maj Thomas E. Griffith, "Improved Combat Search and Rescue," unpublished position paper procured from Lt Col John Blumentritt (Maxwell AFB, AL: Air Command and Staff College, 15 October 1991), 2.
47. Capt John Walsh, interview by the author, 6 October 2000.
48. Gordon and Trainor, *Generals' War*, 263.
49. Col George Gray, interview by the author, 3 May 2001; and Orrell, interview.
50. Orrell, interview.
51. Ibid.
52. Capt Steve Otto, interview by the author, 30 April 2001.
53. Gray, interview.
54. Richard Whittle, "Missions Keep Choppers Flying in Face of Danger," *Dallas Morning News*, 8 March 2002.
55. Ibid.
56. Orrell, interview.
57. Ibid.
58. Ibid.
59. Clancy with Horner, *Every Man a Tiger*, 346.
60. Col James H. Kyle, USAF, retired, with John R. Edison, *The Guts to Try: The Untold Story of the Iran Hostage Rescue Mission by the On-Scene Desert Commander* (New York: Orion Books, 1990), 90; and History of the Air Force Special Operations Command (Hurlburt Field, FL: [hereafter AFSOC history], 1 January 1990–31 December 1991), 86.
61. Kyle with Edison, *Guts to Try*, 96.
62. LT Jeff Zaun, interview by the author, 17 March 2002.
63. Andrews, *Airpower against an Army*, 35.
64. Glosson, interview. Numerous of the AFSOF helicopter pilots were shocked at this statement. As one responded, "Amongst the fighter guys,

there is an attitude of 'die to prove you are brave . . . and stupid;'" Comments by Maj Gen Rich Comer on this book, 2 February 2003.

65. Ibid.
66. Lt Col Joe Hampton, interview by the author, 12 March 2000.
67. Comer, interview.
68. Steube, interview; and Hampton, interview.
69. Comments by Maj Gen Richard Comer on this book, 2 February 2003.
70. Steube, interview.
71. Hampton, interview.
72. Scudder, interview.
73. Glosson, interview.
74. Ibid.
75. Ibid.
76. Ibid.
77. Ibid.
78. Clancy with Horner, *Every Man a Tiger*, 394.
79. Ibid., 410.

Appendix

Comparison of British and US Officer Ranks

NATO/US rank code	British Army & Royal Marines	US Army/ USAF & USMC	Royal Air Force (RAF)	Royal Navy	US Navy
OF-9/O10	General **(Gen)**	General **(GEN/Gen)**	Air Chief Marshal **(ACM)**	Admiral **(Adm)**	Admiral **(ADM)**
OF-8/O9	Lieutenant-General **(Lt Gen)**	Lieutenant-General **(LG/Lt Gen)**	Air Marshal **(AM)**	Vice Admiral **(VADM)**	Vice-Admiral **(VAdm)**
OF-7/O8	Major-General **(Maj Gen)**	Major General **(MG/Maj Gen)**	Air Vice-Marshal **(AVM)**	Rear-Admiral **(RAdm)**	Rear Admiral (Upper Half) **(RADM)**
OF-6/O7	Brigadier **(Brig)**	Brigadier General **(BG/Brig Gen)**	Air Commodore **(ACdre)**	Commodore **(Cdre)**	Rear Admiral (Lower Half) **(RDML)**
OF-5/O6	Colonel **(Col)**	Colonel **(COL/Col)**	Group Captain **(Gp Capt)**	Captain **(Capt)**	Captain **(CAPT)**
OF-4/O5	Lieutenant Colonel **(Lt Col)**	Lieutenant-Colonel **(LTC/Lt Col)**	Wing Commander **(Wg Cdr)**	Commander **(Cdr)**	Commander **(CDR)**
OF-3/O4	Major **(Maj)**	Major **(MAJ/Maj)**	Squadron Leader **(Sqn Ldr)**	Lieutenant Commander **(Lt Cdr)**	Lieutenant Commander **(LCDR)**
OF-2/O3	Captain **(Capt)**	Captain **(CPT/Capt)**	Flight Lieutenant **(Flt Lt)**	Lieutenant **(Lt)**	Lieutenant **(LT)**
OF-1/ O1	Lieutenant **(Lt)**	First Lieutenant **(1LT/1st Lt)**	Flying Officer **(FO)**	Sub Lieutenant **(SLt)**	Lieutenant, Junior Grade **(LTJG)**
O2	Second Lieutenant **(2Lt)**	Second Lieutenant **(2LT/2d Lt)**	Pilot Officer **(PO)**		Ensign **(ENS)**

Note: Some services may use additional or alternate abbreviations. Those above are used throughout this publication for standardization and clarity.

Abbreviations

ABCCC	airborne battlefield command and control center
ACOC	air combat operation center
ACR	armored cavalry regiment
AFRCC	Air Force rescue coordination center
AFSOC	Air Force Special Operations Command
ALO	air liaison officer
ARCENT	Army component to CENTCOM
ARRG	Aerospace Rescue and Recovery Group
ARRS	Aerospace Rescue and Recovery Service
ARRSq	Aerospace Rescue and Recovery Squadron
ARS	Air Rescue Service
ASOC	air support operations center
ATO	air tasking order
AWACS	airborne warning and control system
BCE	battlefield coordination element
CAFMS	computer-assisted force management system
CALCM	conventional air-launched cruise missiles
CAS	close air support
CBU	cluster bomb unit
CENTAF	Air Force component to CENTCOM
CENTCOM	United States Central Command
CEP	circular error probable, a statistical measure of weapon accuracy
CINC	commander in chief of a unified or specified command; a combatant commander
CSAR	combat search and rescue
DALS	downed airman location system
DASC	direct air support center
DOC	designed operational capability
ELF	electronic location finder
EUCOM	European Command
FAF	French Air Force
FAV	fast attack vehicle
FOL	forward operating location
FSCL	fire support coordination line

ABBREVIATIONS

GPS	global positioning system
HARM	high-speed antiradiation missiles
HASC	House Armed Services Committee
HMA	Helicopter Attack Squadron, USMC
HMM	Helicopter Medium Squadron, USMC
INS	inertial navigation system
IRCM	infrared countermeasures
JFACC	joint forces air component commander
JFC-E	Joint Forces Command East
JFC-N	Joint Forces Command North
JOC	joint operations center
JSOACC	joint special operations air component commander
JSOC	Joint Special Operations Command
JSOTF	joint special operations task force
JTF	joint task force
KFIA	King Fahd International Airport
KKMC	King Khalid Military City
KTO	Kuwaiti theater of operations
MAC	Military Airlift Command
MAG	Marine air group
MARCENT	Marine component to CENTCOM
MEF	Marine expeditionary force
MLRS	Multiple-launch rocket system
mm	millimeter, a designation for the size of a weapon
MTACC	Marine tactical air control center
NASA	National Aeronautics and Space Administration
NAVCENT	Navy component to CENTCOM
NRCC	Naval rescue coordination center
NSOA	National Special Operations Agency
NVG	night-vision goggles
OPCON	operational control
OPLAN	operational plan
POL	petroleum, oil, and lubricants

RAF	Royal Air Force
RCC	rescue coordination center
RSAF	Royal Saudi Air Force
RWR	radar warning receiver
SAM	surface-to-air missile
SAR	search and rescue
SARSAT	search and rescue satellites
SARTF	SAR task force
SLGRS	small, lightweight GPS receivers
SOCCENT	Special Operations Command component to CENTCOM
SOCEUR	Special Operations Command component of EUCOM
SOCOM	Special Operations Command
SOF	Special Operations Forces
SOS	special operations squadron
SOW	special operations wing
SPACECOM	United States Space Command
SPINS	special instructions
STG	special tactics group
STS	special tactics squadron
STU	secure telephone unit
TAC	Tactical Air Command
TACC	tactical air control center
TACON	tactical control
TALD	tactical air-launched decoy
TFS	tactical fighter squadron
TFW	tactical fighter wing
TOC	tactical operations center
TPFDL	time-phased force and deployment list
TRAP	tactical recovery of aircraft and personnel
TSAR	theater search and rescue
VA	Attack Squadron, US Navy
VAQ	Electronic Warfare Squadron, US Navy
VF/A	Fighter/Attack Squadron, US Navy
VMA	Attack Squadron, USMC
VMF/A	Fighter/Attack Squadron, USMC

ABBREVIATIONS

WGS	World Geodetic System—the reference frame defined by the National Imagery and Mapping Agency and used by DOD for all its mapping, charting, surveying, and navigation needs, including its GPS "broadcast" and "precise" orbits
WSO	weapons system officer

Bibliography

Acree, Cynthia B., with Lt Col Cliff M. Acree. *The Gulf between Us: Love and Terror in Desert Storm.* Washington, DC: Brassey's Press, 2000.

Adams, Thomas K. *US Special Operations Forces in Action: The Challenge of Unconventional Warfare.* Portland, OR: Frank Cass Press, 1998.

Aerospace Rescue and Recovery Service 1946–1981, An Illustrated Chronology. Military Airlift Command History, 1983.

Air Force Doctrine Document (AFDD) 1. *Air Force Basic Doctrine,* 1 September 1997.

Albright, Joseph. "Two Myrtle Beach Pilots Help Rescue Downed Flyer." *Strand Sentry,* 25 January 1991, 11.

Allen, Charles. *Thunder and Lightning: The RAF in the Gulf: Personal Experiences of War.* London, UK: Her Majesty's Stationery Office, 1991.

Andrews, Capt Bill. Mission map for Mutt 41, n.p., n.d.

Andrews, Lt Col William F. *Airpower against an Army.* Maxwell AFB, AL: Air University Press, 1998.

Association of the US Army. *The U.S. Army in Operation Desert Storm.* Arlington, VA, June 1991.

Atkinson, Rick. *Crusade: The Untold Story of the Persian Gulf War.* Boston, MA: Houghton Mifflin, 1993.

Baldwin, Sherman. *Ironclaw: A Navy Pilot's Gulf War Experience.* New York: William Morrow, 1996.

Balman, Sid. "Second: U.S. Force Planned to Invade Tehran to Free 52." *Air Force Times,* 25 September 1989, 24.

Blumentritt, Maj John W. *Playing Defense and Offense: Employing Rescue Resources as Offensive Weapons.* Maxwell AFB, AL: School of Advanced Airpower Studies, Air University, May 1999.

Booher, 1st Lt Joey. "Gulf War Journal." *Code One Magazine,* April 1992, 20.

Bradin, James W. *From Hot Air to Hellfire: The History of Army Attack Aviation.* Novato, CA: Presidio Press, 1994.

Breuninger, Michael S. *United States Combat Aircrew Survival Equipment: World War II to the Present: a Reference Guide for Collectors.* Atglen, PA: Schiffer Publishers, 1995.

Canan, James W. "The Electronic Storm." *Air Force Magazine*, June 1991.

Carpenter, Maj Mason. *Joint Operations in the Gulf War*. Maxwell AFB, AL: School of Advanced Airpower Studies, Air University, February 1995.

Center for Naval Analysis (CNA). "Desert Storm Reconstruction Report, Volume II: Strike Warfare." Arlington, VA: October 1991.

Chinnery, Philip D. *Any Time, Any Place: Fifty years of the USAF Air Commando and Special Operations Forces, 1944–1994*. Annapolis, MD: Naval Institute Press, 1994.

Clancy, Tom, with Gen Carl Stiner. *Shadow Warriors*. New York: G. P. Putnam's Sons, 2002.

Clancy, Tom, with Gen Chuck Horner. *Every Man a Tiger*. New York: G. P. Putnam's Sons, 1999.

Collins, SSgt Patricia. "Awards Recognize 20th SOS Bravery." *Commando*, 20 September 1991.

Command Histories of US Navy ships and aviation squadrons. Washington, DC: Naval Historical Center, n.d.

Cooper, Dale B. "Bulldog Balwanz and His Eight Man Army." *Soldier of Fortune*, May 1992.

———. "Young Guns, Harrier Pilots in the Persian Gulf." *Soldier of Fortune*, May 1991, 52.

Cooper, TSgt Ray. 1723d Special Tactics Squadron: Alert Log notes for 19–20 January 1991, Stroke 65 Mission.

Cornum, Rhonda, with Peter Copeland. *She Went to War: The Rhonda Cornum Story*. Novato, CA: Presidio Press, 1992.

Cureton, Lt Col Charles H. *U.S. Marines in the Persian Gulf, 1990–1991: with the 1st Marine Division in Desert Shield and Desert Storm*. Washington, DC: History and Museums Division, Headquarters US Marine Corps, 1993.

Davis, Richard G. *The 31 Initiatives: A Study in Air Force–Army Cooperation*. Washington, DC: Office of Air Force History, 1987.

de la Billière, Gen Sir Peter. *Storm Command: A Personal Account of the Gulf War*. New York: Harper Collins Publishers, 1992.

Desert Shield/Desert Storm: Air Force Special Operations Command (AFSOC) in the Gulf War. Hurlburt Field, FL: History Office, Air Force Special Operations Command, 1992.

"Desert Storm, Final Report to Congress: Conduct of the Persian Gulf War, 1991." *Military History Magazine*, April 1992.

Dorr, Robert. "POWs in Iraq Survived Thanks to Training, Courage, Faith." *Naval Aviation News*, May–June 1991, 6.

Every, Martin G. *Navy Combat Search and Rescue*. Falls Church, VA: BioTechnology Inc., September 1979.

Falzone, Maj Joseph J. *Combat Search and Rescue CSEL Enhancements for Winning Air Campaigns*. Maxwell AFB, AL: Air University Press, December 1994.

Flanagan, Edward M., Jr. "Hostile Territory was Their AO in Desert Storm." *Army*, September 1991, 12.

———. *Lightning: The 101st in the Gulf War*. Washington, DC: Brassey's Press, 1993.

Fuqua, CAPT Michael, USN. "We Can Fix SAR in the Navy." *Proceedings*, September 1997, 57.

Glosson, Lt Gen Buster C. *War with Iraq: Critical Lessons*. Charlotte, NC: Glosson Family Foundation, 2003.

Gordon, Michael R., and Gen Bernard E. Trainor. *The Generals' War: The Inside Story of the Conflict in the Gulf*. New York: Little, Brown and Co., 1995.

Griffith, Maj Thomas E. "Improved Combat Search and Rescue." Unpublished position paper. Maxwell AFB, AL: Air Command and Staff College, procured from Lt Col John Blumentritt, 15 October 1991.

Guilmartin, John F., Jr. *A Very Short War: The Mayaguez and the Battle of Koh Tang*. College Station, TX: Texas A&M University Press, 1995.

Gulf War Air Power Survey (GWAPS) Summary Report. Washington, DC: Government Printing Office (GPO), 1993.

Hales, Grant. *Airpower in Desert Storm: Iraq's POWs Speak*. Draft, n.d.

———. *Desert Shield/Desert Storm Chronology*. Headquarters Tactical Air Command, 26 June 1991.

Hallion, Richard P. *Storm over Iraq*. Washington, DC: Smithsonian Institution Press, 1992.

Hampton, Capt Dan. "The Weasels at War." *Air Force Magazine*, July 1991.

Herlick, Edward. "Daring Rescue Deep in Iraq." *Military History*, December 1994, 62.

———. *Separated by War: An Oral History by Desert Storm Fliers and Their Families.* Blue Ridge Summit, PA: Tab Aero Press, 1994.

Herman, Lt Col Theodore, USMC, retired. "Harriers in the Breach." *Proceedings*, February 1996, 44.

History of the Twenty-third Air Force and the Aerospace Rescue and Recovery Service, 1984–1985 Chronology, n.p., n.d.

Hobson, Chris. *Vietnam Air Losses.* Hinkley, UK: Midland Publishing, 2001.

Hunt, Peter. *Angles of Attack, an A-6 Intruder Pilot's War.* New York: Ballantine Books, 2002.

International Civil Aeronautical Organization Circular 185-AN/121. *Satellite Aided Search and Rescue—The COPAS SARSAT System.* Montreal, Canada: International Civil Aeronautical Organization, 1986.

Jamieson, Perry D. *Lucrative Targets: The U.S. Air Force in the Kuwaiti Theater of Operations.* Washington, DC: Air Force History and Museums Program, GPO, 2001.

Joint Chiefs of Staff. Joint Review U.S. POW/MIA and EPW Issues, June 1991.

Joint Publication 1-02, *Department of Defense Dictionary of Military and Associated Terms.* 2001.

Kamiya, MAJ Jason K. *A History of the 24th Mechanized Infantry Division Combat Team during Operation Desert Storm: The Attack to Free Kuwait (January–March 1991).* Fort Stewart, GA: Headquarters 24th Mechanized Infantry Division, 1992.

Keane, Kathleen. "Medal Puts Pilot in Limelight." *Strand Sentry*, 16 August 1991, 1.

Keaney, Thomas A., and Eliot A. Cohen. *Gulf War Air Power Survey (GWAPS).* Washington, DC: GPO, 1993.

Khalifouh, Col Ali Abdul-Lateef. *Kuwaiti Resistance as Revealed by Iraqi Documents.* Mansouria, Kuwait: Center for Research and Studies on Kuwait, 1994.

Kinnear, CDR Neil, USNR. "HCS 4 Remembers 'Spike' Det." *Rudder*, March–April 1991, 4.

Kyle, Col James H., USAF, retired, with John R. Edison. *The Guts to Try: The Untold Story of the Iran Hostage Rescue*

Mission by the On-Scene Desert Commander. New York: Orion Books, 1990.

LaRue, Steve. "Marine Missing, Feared First San Diego Casualty." *San Diego Union*, 20 January 1991, A–9.

LeMenager, Capt William. "A Gulf War Chronicle," HQ SOCOM, unpublished personal manuscript, January 1998.

Leyden, Andrew. *Gulf War Debriefing Book: An After Action Report.* Grants Pass, OR: Hellgate Press, 1997.

Mall, Maj Gen William J. "Commander Shares Insights." *Airlift Magazine*, Fall 1984, 1–3.

Marolda, Edward, and Robert Schneller Jr. *Shield and Sword.* Washington, DC: Naval Historical Center, Department of the Navy, 1998.

Marquis, Susan L. *Unconventional Warfare.* Washington, DC: Brookings Institution Press, 1997.

McNab, Andy. *Bravo Two Zero.* New York: Island Books, 1993.

Micheletti, Eric. *Air War over the Gulf.* London: Windrow and Greene, 1991.

Morse, Stan, ed. "Gulf Air War Debrief, (GAWD)." *World Air Power Journal*, 1991.

Murray, Williamson, with Wayne Thompson. *Air War in the Persian Gulf.* Baltimore, MD: Nautical and Aviation Publishing Company of America, 1996.

Nadel, Joel, and J. R. Wright. *Special Men and Special Missions: Inside American Special Operations Forces, 1945 to the Present.* London: Greenhill Books, 1994.

"Navy Changes Status of A-6 Pilot from Missing to Killed in Action." *Star-Ledger*, 9 May 1991.

O'Boyle, Capt Randy. CSAR map for Iraq, n.p., n.d.

O'Brien, Dennis. "From POW to Pilot Again." *Virginian-Pilot*, 8 February 2003, A–1.

Peters, John, and John Nichol. *Tornado Down.* London: Penguin Books, 1993.

Pokrant, Marvin. *Desert Storm at Sea: What the Navy Really Did.* Westport, CT: Greenwood Press, 1999.

Quilter, Col Charles J., II. *U.S. Marines in the Persian Gulf, 1990–1991: With the Marine Expeditionary Force in Desert Shield and Desert Storm.* Washington, DC: History and Museums Division Headquarters, US Marine Corps, 1993.

Rip, Michael R., and James M. Hasik. *The Precision Revolution: GPS and the Future of Aerial Warfare.* Annapolis, MD: Naval Institute Press, 2002.

Ripley, Tim. "Desert Weasels." *United States Air Force Yearbook 1992*, 60.

Robison, LT Russ, USNR. "Reserve SAR Unit Home from Desert Storm." *Free Navy News*, 27 March 1991, 1.

Ryan, Chris. *The One That Got Away.* London, UK: Century Press, 1995.

Santoli, Al. *Leading the Way: How Vietnam Veterans Rebuilt the U.S. Military: An Oral History.* New York: Ballantine Books, 1993.

Scales, BG Robert H. *Certain Victory: The US Army in the Gulf War.* Washington, DC: Office of the Chief of Staff, US Army, 1993.

Schemmer, Benjamin. "No USAF Combat Rescue in Gulf: It Took 72 Hours to Launch One Rescue." *Armed Forces International*, July 1991.

Schwartz, John. "Bringing Them Back Alive." *Newsweek*, 4 February 1991.

Schwarzkopf, GEN H. Norman, with Peter Petre. *It Doesn't Take a Hero.* New York: Bantam Books, 1992.

Sharrett, CDR Patrick, USN; Lt Col Joseph Wysocki, Capt Gary Freeland, USAF; Capt Scott Netherland, US Army; and Donald Brown. "GPS Performance: An Initial Assessment," *Navigation Journal* 39, no. 1 (Spring 1992): 411.

Sherwood, John D. *Fast Movers: America's Jet Pilots and the Vietnam Experience.* New York: Free Press, 1999.

Singley, LT John, USN. "F-14 Crew Reunited after Being Shot Down over Iraq." *Super Sara Review*, USS *Saratoga*, February 1991, 8.

Smallwood, William L. *Strike Eagle: Flying the F-15E in the Gulf War.* Washington, DC: Brassey's Press, 1994.

———. *Warthog: Flying the A-10 in the Gulf War.* Washington, DC: Brassey's Press, 1993.

Smith, Barry D. "Tim Bennett's War." *Air Force Magazine*, January 1993.

Smith, Lt Col Clyde, USMC, retired. "That Others May Live." *Proceedings*, April 1996, 82–88.

Smith, MAJ Kevin. *United States Army Aviation during Operations Desert Shield and Desert Storm.* Fort Rucker, AL: US Army Aviation Center, 1993.

Sochurek, Howard. "Air Rescue behind Enemy Lines." *National Geographic Magazine,* September 1968, 68.

Stearns, Lt Col LeRoy. *U.S. Marines in the Persian Gulf, 1990–1991: The 3d Marine Aircraft Wing in Desert Shield and Desert Storm.* Washington, DC: History and Museums Division, Headquarters US Marine Corps, 1999.

Stout, Jay A. *Hornets over Kuwait.* Annapolis, MD: Naval Institute Press, 1997.

Taylor, Thomas. *Lightning in the Storm, The 101st Air Assault Division in the Gulf War.* New York: Hippocrene Books, 1994.

Tilford, Earl H., Jr. *The United States Air Force Search and Rescue in Southeast Asia, 1961–1975.* Washington, DC: Office of Air Force History, 1980.

Tirpak, John. "The Secret Squirrels." *Air Force Magazine,* April 1994.

Tyner, Lt Col Joe E. *AF Rescue and AFSOF: Overcoming Past Rivalries for Combat Rescue Partnership Tomorrow.* Maxwell AFB, AL: National Defense Fellows Program, Headquarters USAF and Air University, n.d.

United States Special Operations Command History (SOCOM History). MacDill AFB, FL, November 1999.

Vriesenga, Capt Michael P., ed. *From the Line in the Sand: Accounts of USAF Company Grade Officers in Support of Desert Shield/Desert Storm.* Maxwell AFB, AL: Air University Press, 1994.

Waller, Douglas. "Exclusive: Behind Enemy Lines." *Newsweek,* 28 October 1991, 34.

Welsh, Lt Col Mark A. "Day of the Killer Scouts." *Air Force Magazine,* April 1993.

Wetterhahn, Ralph. *The Last Battle: The* Mayaguez *Incident and the End of the Vietnam War.* New York: Carroll and Graf Publishers, 2001.

Whitcomb, Darrel D. *The Rescue of Bat 21.* Annapolis, MD: Naval Institute Press, 1998.

Whittle, Richard. "Missions Keep Choppers Flying in Face of Danger." *Dallas Morning News,* 8 March 2002.

Wilson, George C. "Death Trap in Iraq." *Army Times*, 5 February 1996.

Yarsinske, Amy W. *No One Left Behind: The Lt. Comdr. Michael Scott Speicher Story.* New York: Dutton Books, 2002.

Index

1st Armored Division, 174, 219, 222–24
1st Battalion of the 101st Aviation Regiment, 71, 229
1st Battalion of the 227th Aviation Regiment, 220
1st Battalion of the 5th Special Forces Group, 202
1st Battalion of the 7th Marines, 216
1st Cavalry Division, 219–20
1st Marine Division, 209, 215, 217, 219, 242
1st Mechanized Infantry Division, 214
1st Special Operations Wing, 18–19, 21, 30, 46, 48, 50, 53, 76, 79, 127, 140

2d Air Division Staff, 27
2d Armored Cavalry Regiment, 219
2d Battalion of the 229th Aviation Regiment (2/229), 230
2d Battalion of the 5th Special Forces, 65
2d Marine Division, 209, 213, 216, 218
2d Satellite Control Squadron, 66, 248
2d/7th Infantry Battalion, 236, 241

3d Air Rescue and Recovery Group, 15
3d Armored Cavalry Regiment, 213
3d Armored Division (Iraqi), 173–76, 213
3d Battalion of 160th Special Operations Aviation Regiment, 65, 95, 117, 229
3d Marine Air Wing, 188, 208, 214, 242

4th Tactical Fighter Squadron, 181

6th Light Armored Division (French), 213
6th Squadron, Royal Air Force, 186

7 Squadron, Royal Air Force, 171

8th Special Operations Squadron, 46, 57

9th Special Operations Squadron, 46, 52

10th Special Forces Group, 74
10th Tactical Fighter Squadron, 82, 128, 206, 224, 234

17th Armored Division (Iraqi), 184, 193

20th Special Operations Squadron, 21–22, 31–32, 46, 48, 51–52, 58, 62, 64–65, 70, 73–75, 90–91, 116, 171, 184, 257, 262–63
21st Special Operations Squadron, 85, 264
23 mm guns, 5
23d Tactical Air Support Squadron, 24
24th Mechanized Infantry Division, 213
30 mm fire/guns/cannons, 92, 151, 213, 224
37 mm guns, 5
39th Special Operations Wing, 74
40th Air Rescue and Recovery Squadron, 1, 9, 15–16, 31
48th Tactical Fighter Wing (TFW), 155
52d Tactical Fighter Wing, 73, 112
55th Air Rescue and Recovery Squadron, 24
55th Special Operations Squadron, 46, 58, 68, 91, 143, 175, 179, 185
56th Special Operations Wing, 1, 13
71st Air Rescue Squadron, 47
71st Special Operations Squadron, 65, 109, 179
101st Airborne Division, 71, 213, 221, 224, 228
353d Special Operations Wing, 30, 35
390th Electronic Combat Squadron, 92
401st Tactical Fighter Wing, 113
614th Tactical Fighter Squadron, 114, 118, 157–58
706th Tactical Fighter Squadron AF Reserve unit, 81
1551st Air Rescue and Recovery Squadron (ARRS), 18, 20
1720th Special Tactics Group, 36, 46, 95
1723d Special Tactics Squadron, 58, 162
1730th Pararescue Squadron, 36

INDEX

1986 Congressional Act, (Goldwater-Nichols), 29, 34, 48

1987 Defense Authorization Bill, 29

7440th Composite Wing (Provisional), 85

A-1 Douglas Skyraiders, 2

A-6, 1, 87, 97, 106, 109–10, 142–43, 161, 178, 211, 262

A-7, 27

A-10, 27, 63, 81, 146–47, 150, 158, 167, 178, 186, 189, 194–95, 200, 208, 261

AAA, antiaircraft artillery, 11, 69, 72, 96–97, 100, 102–7, 110–12, 114, 117, 119–20, 124, 128, 131, 133, 135–36, 144, 155, 157, 171–72, 178, 182–83, 191, 210, 212, 217–18, 223, 225, 240, 247, 249–50, 267

Aberg, John, 197

AC-130 gunship, Spectre (See also Thug), 10, 174

ACOC, air combat operation center, 61

Acree, Cliff, 108, 161

Active 304/F-18, USN (I040), 169, 263

ad hoc task force, 110, 153

aerial wheel, 7

AFRCC, Air Force Rescue Coordination Center, 15, 34, 57

AFSOC, Air Force Special Operations Command, 31–32, 36–39, 46–50, 57–58, 65, 76–78, 134, 138, 140, 159–60, 164, 251–56, 258, 266–67, 269, 272, 274

AFSOCCENT, AF component of SOCCENT, 45–46, 53, 263

AG4 kill box, 178

AH-1, Army/Marine Cobra helicopter, 9, 108, 210, 215

AH-64, Apache helicopter, 70, 91, 220, 228, 230–31, 233, 236, 239, 259, 263

air campaign plan, 52, 54, 63, 67, 69, 82, 108, 121, 252, 254

Air Force Council, 25, 27

Air Force Cross, 3, 6, 8, 12, 46–47, 243

Air Force RCC, 60, 269

air liaison officer, 61, 213, 224

Air National Guard, 18, 35, 49

Air Staff, 9, 21, 23–25, 27, 31, 33–34, 47, 49, 52–53, 83, 274

air superiority/supremacy, 67, 114, 154–55, 168, 172, 249, 252, 264, 266–67

AK-47s, 62, 109, 204, 225, 266

Al Ahsa Air Base, Saudi Arabia, 61, 102

Al Asad Air Base, 142, 144

Al Busayyah, Iraq, 219

Al Firdos bunker, 185, 191

Al Hammar bog, 230

al Jabayl, Saudi Arabia, 169

Al Jaber Airfield, Kuwait, 101, 215

Al Jahra, Kuwait, 218

Al Kharj Air Base, 122

Al Qāim, 128, 130–36, 155, 171, 183

Al Tallil Airfield, 120

Al Taqaddum, 96, 104, 191, 247

Alaska, 35, 47

Ali Al Salem Airfield, Kuwait, 98

Al-Jouf, Saudi Arabia, 62, 64

ALO (air liaison officers), 61, 213, 224

ALQ-119/131/184/135 jamming pods, 169, 249

Amman, Jordan, 127

Andrews, Bill, 82, 158, 205–6, 225–31, 234, 243, 265, 267, 273–74

Ankerson, Robert, 167

antiradar jamming, 105

antiradiation missile, 5

AOR (area of responsibility), 54–57, 59, 62, 90, 141

Ar Rumaylah Airfield, 100, 167

Ar Rutbah radar site, 156

ArAr, Saudi Arabia, 62, 64, 75, 86, 89, 93, 116, 126, 136, 143–44, 146, 148, 171, 176, 186, 209

ARCENT, US Army Central Command, 45, 60, 208, 243

Army RCC, 60, 169–70, 183, 195, 199, 201, 223–24, 239

Army Special Forces teams/troops, 50, 68, 75, 134, 198, 263

ARRS (Aerospace Rescue and Recovery Service), 15–18, 22–25, 27–28, 31, 34, 36, 40

ARS (Air Rescue Service), 20–21, 34–36, 39, 41, 46–47, 49, 57, 77, 252–53, 270

Arthur, Stanley, 111, 175

As Shuaybah, port of, 218

ATO (air tasking order), 85, 136

AV-8, 157, 172, 184, 211, 216–18, 223, 261

292

INDEX

Avenger 11/AH-64, US Army (I088), 239
AWACS (airborne warning and control system), 38–39, 63, 85, 89, 94, 97, 101–2, 104–6, 115–17, 119, 125, 128, 142–43, 145–50, 152, 157, 171, 175, 179, 187, 194, 196–99, 204, 226, 233–36, 248, 256
Az Zubayr River, Iraq, 109

B-52, 87, 94, 104, 113, 180, 257
Baghdad Billy, 187–88
Baghdad, 70–71, 85, 93–94, 96, 113–14, 119, 121–22, 128, 136, 142, 144, 149, 185, 187–88, 191, 195, 199, 218, 240, 247, 250, 268
Baker, James, 44
Baldwin, Sherman, 88, 159, 189
Balkans, 272
Ball, Jon, 157–158, 186, 222
Balwanz, Richard "Bulldog," 170, 202–8
Barksdale AFB, LA, 98
Barretta 11, 200
Basra, Iraq, 104, 178
Batman Air Base, Turkey, 74
Batson, Peter, 141
Battle Force Yankee, (aka: CTF 155), 61, 64
Battle Force Zulu, (aka: CTF 151), 60, 109
Battle of Khafji, 176
BCE (battlefield coordination element), 180, 228–29
Beard, Mike, 179
Bedouin camps, 95, 132
Bedouin tribesmen, 116
Behery, Ahmed, 59
Beirut, Lebanon, 18
Belfast 41/Tornado, RAF (I072), 191
Bengal 15/UH-60, US Army (I087/M020), 224, 230, 232–34, 236–38, 241, 259, 265, 271
Bengal 505, 1, 10, 46, 139
Benji 53/F-16, USAF (I078/M017), 196, 198, 202, 228, 259, 263–64
Bergan 23/A-4, Kuwaiti Air Force (I010), 102, 259
Berryman, Michael, 172
BGM-109 Tomahawk cruise missiles, 71
BioTechnology Inc., 9, 12
Bioty, John, 161, 184, 242
Black Hawks, UH-60s, 169, 195, 223–24, 231–32, 259, 267

Black Hole, 52–54, 67, 69, 86, 122
Black, Ken, 57
Blackwood, Jim, 227, 243
Block 40 F-16C/Ds, 39, 139, 248, 256
blood chit, 132
Blue Flag, 252
Boomer, Walter, 218
Boorda, Jeremy, 100
Bosnia, 76
BQM-74 unmanned drone aircraft, 71
Bradt, Douglas, 186
Bravo 10, 132, 155
Bravo 20, 155, 171–72, 255, 265
Bravo 30, 155
Brazen, British warship, 174
Brelinski, Roger, 237
Brinson, Bill, 5, 12
British forces, 75, 140, 155
Bryan, Bill, 205, 230–31, 233–34, 237–38, 241, 244
Buccaneers, 169, 191
Burgan oil field, 215
Burshnick, Tony, 25, 41
Bush, George H. W., 43, 64, 69, 85, 87, 107, 167, 240, 265
Butler, Mike, 220
Butts, William, 234, 237

C-130, 18, 68, 140
C-141, 50
C-17, 25
C-5, 17, 50
Caesar 44/Tornado, Italian Air Force (I015), 107
CALCM (conventional air-launched cruise missiles), 98
CAP, combat air patrol, 117, 145, 247
Carlton, P. K., 16–17
Carter, Jimmy, 19, 21–22
Cassidy, Duane, 27, 34
Cat 36/AV-8, USMC (I047/M009), 172, 258
CBU-87 cluster bombs, 205
CENTAF (Central Command Air Force), 45, 51–53, 57, 67, 86, 89, 116, 129, 138–42, 153–54, 158, 170, 176–177, 187, 189, 212, 254–55, 267, 272
CENTCOM (Central Command), 43
CEP (circular error probable), 121, 248
CH-3, 18
CH-46, 201
CH-47, 21, 155, 171

293

INDEX

CH-53, 11
chaff (radar confusion reflectors), 105, 120, 172, 187
Checkmate division, 52
Cheney and McKay Trophy awards, 15
Cheney, Richard "Dick," 15, 43–44, 99–100, 185
Chevy 06, 129
chief of staff, 9, 21, 23, 31, 34, 47, 164, 189
Chinook, *See* CH-47
Chowder 21, 224
CIA, 28
CINC (commander in chief), 28–29, 122, 271
Clap 74/F-16, USAF (I024), 118, 121
Clark, Rupert, 193, 261
Clem, J. D., 131, 133–34, 163
close air support, 27, 201, 246, 250
CNN (Cable News Network), 136–37, 195
Cobras, *See* AH-1s
Cody, Dick, 71, 229, 231
Cohen, Bill, 29, 160, 188, 273
Cold Lake, Alberta, Canada, 252
Colier, Robert, 104
Combat 13, 12, 216–17
Combined Task Force, 60
Comer, Rich, 30–32, 41–42, 48–49, 51–52, 58–59, 64, 68, 70–71, 73, 76–79, 82, 90, 116, 118, 152, 155, 162, 164, 171–72, 184, 188–89, 244, 257, 269, 271, 274–75
COMSOCCENT, 55
COMUSARCENT, 55
COMUSCENTAF, 55, 57, 77, 267
COMUSMARCENT, 55
COMUSNAVCENT, 55
Connor, Pat, 178
Cooke, Barry, 178
Cornum, Rhonda, 233–34, 237, 244
Corvette 03/F-15E, USAF (I028/M007), 103, 122–25, 128, 131–33, 135–37, 139, 141, 163, 187, 254–58, 262, 269–70
Costen, William, 110
Cox, Bruce, 115–16
CR (combat rescue), 61
Crandall, Richard, 130
crowd control weapons, 2
cruise missiles, 70–71, 89–90, 98, 119, 195, 248
CSAR plan, 44, 53–54, 57, 59, 67, 86, 110, 126
CSAR (combat search and rescue), 3, 9–10, 12, 25, 27–29, 33, 36, 39–40, 44, 47, 53–55, 57–62, 64, 67–68, 74–77, 79, 82–86, 89–90, 99, 102, 110, 126, 129, 138, 141, 145, 147, 149, 152–53, 155, 158, 164, 170, 176–77, 184, 196, 198, 202, 205, 211, 226, 232, 234–35, 239, 245, 247, 252–59, 262, 264, 267–72, 274
CSAREXes (CSAR exercises), 67, 254
CTF 155, Red Sea Battle Group, 64
Cyclops 47 and 48/2 OV-1s, US Army (I089), 239
Daily, Dell, 15, 20, 33, 51, 85, 99, 136, 197, 229
daisy-chain flying pattern, 150–51
DALS (downed airman location system), 23, 37–38, 61, 63, 139, 256
Daniel, Dan, 26–29, 237
Dark 13/Tornado, RSAF (I026/M004), 13, 119
Davis-Monthan AFB, AZ, 65
Day, Otis (Pointer 75), 235, 238
de la Billière, Peter, 140, 164
Decuir, Slammer, 137
Delta Force commandos, 209
Demers, Paul (Pointer 75), 235, 238
Department of Defense, 78, 164
Deptula, Dave, 53–54, 86
Deputy Chief of Staff for Plans and Operations, 23
Desert Hawk 102/UH-60, RSAF (I042), 102, 169
Desert One, 19, 21, 28, 49, 81, 92, 161, 249, 258, 266, 270
DF (direction finder/ing), 97, 117–18, 130
Dhahban, Oman, 173
Dhahran, Saudi Arabia, 50
Diego Garcia, 59, 113, 180
Diyarbakur, Turkey, 200
DO (director of operations), 137
DOC (designed operational capability), 33, 39
Dodson, Eric (Benji 54), 196–97
Doman, Al, 227
Doppler navigation system, 9, 16, 66
Dover 02/Tornado, RAF (I039), 167
drones, 90, 94, 96
dumb bombs, 111
Dunlap, Troy, 237

Durham, Randy, 50, 52, 243
Dustoff 229, 183
Dwyer, Robert, 181
E-2 Hawkeye, US Navy, 109–10, 186
E-8 Joint STARS, 248
EA-6B, 142, 146, 249
east-west highway, from Baghdad to Jordan, 149
Eberly, David, 122–25, 127–28, 130, 132–33, 136–37, 139, 162, 254
Edwards, Jon, 180
EF-111 jammers, 92
Egyptian forces, 214, 219
Eichenlaub, Paul, 186
electronic war, 70, 267
ELF (electronic location finder), 37
Elsdon, Nigel, 104
ELT (emergency location transmitter [beeper activation]), 90
emergency exfiltrations, 202
Enfield 37 and 38/A-10s, USAF (I073/M016), 193
error probable, 38, 121, 248
EUCOM (European Command), 74
Euphrates River/valley, 221
Exercise Desert Force, 68
Exercise Search and Rescue, 68

F/A-18, 169
F-100, 9
F-105, 5
F-111, 98, 187, 251
F-111F, 182
F-117, 89–90, 112, 195, 240
F-14, 87, 132, 142–44, 263
F-15, 94, 117, 127, 186–87, 267
F-15E, 104, 111, 122, 125, 128–30, 163, 208, 254, 256, 262
F-16, 114, 118–19, 157, 196–97, 200, 206, 224–25, 227–28, 232, 234–35, 239–40, 245, 267
F-16C, 39, 139, 248, 256
F-18, 82, 87, 97, 141, 169, 172, 181, 216, 223, 259
F-4G Wild Weasels, 70, 72, 85, 114, 124, 182
F-5, RSAF, 185, 261
factory, the (F-16 geo-reference point), 225
Fairchild AFB, WA, 57
Falstaff 66/F-4G (I021), 111
fast forward air controller (FASTFAC), 182, 216, 223
fast mover aircraft, 172

FAV (fast attack vehicles), 64
Faylaka Island, Persian Gulf, 178
Fitzsimmons, Scott, 159, 222–26, 243
FLIR (forward-looking infrared system), 16, 24, 68, 182
Flood, Al, 71, 232, 244
FLOT (forward line of own troops), 138
FMU-139 electronic fuses, 157–58
FOB (forward operations base), 232
Fogleman, Ronald, 9
FOL (forward operating location), 197, 216
Folse, Dan, 95
Force Package Q, 113, 118
Fort Campbell, KY, 196
Fort Irwin, CA, 27
Fort Rucker, AL, 230, 243, 274
forward air controllers, 2, 4, 38, 109, 171, 181, 213, 215, 222
Forward Look, 27, 108
Fox, Jeff, 199–200
French Air Force, 61, 101–2, 241
FROG, free rocket over ground, 108, 172, 175
FSCL (fire support coordination line), 222, 228
FY84 Appropriations Bill, 25

Gabram, Doug, 230
Gale, Al, 130
Garlington, Jerry, 143, 148
Garvey, Philip (Bengal 15), 232–34, 237
GBU-15, laser-guided bombs, 167
G-day, beginning of the ground attack, 201
Gee, George, 87
Germany, 12, 73, 156, 253
Ghost 02/AC-130, USAF (I033), 153–54
Glosson, Buster, 52, 54, 77, 81, 200, 247, 254, 267, 273
Godfrey, Robert, 232, 234, 237
Goff, Randy, 145, 148
Goldwater-Nichols Act, 29, 34, 48
GPS (global positioning system), 38–39, 42, 44, 66, 70, 76, 78, 91, 113, 139, 149, 153, 162, 198, 202–4, 206–8, 211–12, 242, 245, 248–49, 256–58, 272–74
Gray, George, 13, 45–46, 48, 53, 65, 76, 102, 126, 130–31, 138, 140, 161, 189, 243, 263–66, 268, 274
Green Flag electronic warfare training, 251
Grenada, 26, 28

INDEX

Griffith, Tom, 122–23, 125, 127–30, 132, 136–37, 139, 254, 274
Gulf of Aden, 59
Gulf of Tonkin, 1
Gunfighter 126/UH-1N, USMC (I056), 126, 181

H-2/H-3 airfields, 121, 183
Hafar al Batin, 170
Hampton, Joe, 57–61, 63, 77–78, 85–86, 126, 129–32, 156, 159, 161–64, 180, 189, 198, 208, 226–28, 243, 269, 275
Hanoi, Vietnam, 8, 11, 86, 114
Harden, Grant, 103, 131, 133–36, 161, 163
Harding, Jim (Sandy 01), 4–7
Hardziej, Dennis, 92, 94
HARM (high-speed antiradiation missile), 138, 142, 182, 267
Harmon, Paul, 115–16, 162, 184, 189
Harris, Bill, 2
Harvell, Pete, 44, 48, 50–51, 54, 60, 63, 76–78, 138, 161, 163, 255, 274
Harward, Don, 95
HASC (House Armed Services Committee), 26, 28
HC-130, 2–3, 27, 35, 47, 52, 256
HC-130P/N Combat Shadows, 46
HC-7, US Navy helicopter squadron 7, 1, 36
HC-9, 36
HCS-4, Red Wolves, 64
HCS-4/5, Spikes, 64
HCS-5, Firehawks, 64
Health 67/Dalphin, RSAF (I085), 67, 220
Heartless 531/A-6, USN (I052), 178
Heath, Mike, 141
heat-seeking missile(s), 11, 68, 99, 101–2, 108, 172, 178, 184, 194–95, 212, 214, 216–18, 222–23, 226, 267
Hebein, Mark, 234–35, 238, 244
Helgeson, Larry, 47, 76
Hellfire missiles and rockets, 70
Hendrickson, Dean, 88
Henry, Larry "Poobah," 70–72
HH-3 school, Kirtland, NM, 28
HH-3, 15, 28, 35–36, 47, 65
HH-53 Jolly Greens, 9
HH-60A, 25
HH-60D Nighthawk, 22
HH-60E, 25
HH-60G Pave Hawk, 24

HH-60H Seahawk, 36
Hicks, Stephen, 193
high bird, backup, 4
highway of death, 219
HMS *Brazen*, 174
HMS *Cardiff*, 174
HMS *Gloucester*, 174, 218
Ho Chi Minh Trail, 1, 5, 10, 24, 128, 140
Hodler, Dan, 57
Holland, Donnie, 105
Homan, Mike, 151
home in/homing in, 37, 72, 97, 145, 151, 178, 256
homing missiles, 2
Honey Badger, 21–22
Horner, Charles A. "Chuck," 43, 48, 51–52, 55–56, 58–59, 62, 69–70, 72, 75, 78–79, 81, 85–86, 99, 101, 112, 121–22, 137–38, 154, 159–60, 163–64, 167, 170–71, 174, 189–90, 195, 209, 212, 249–50, 264, 266–69, 271–75
Hostage 75/OV-10, USMC (I016/M002), 108–9, 258
Houle, Ed, 128–29, 131, 163
HSL-44, 157
HUD (heads up display), 157
Hulk 46/B-52, USAF (I055), 180
HUMINT (human intelligence), 202
Hunter 26, 160, 185–86, 258, 261
Hunter, Guy, 108
Hunter, Russ, 95, 160
Hurlburt Field, FL, 76–77, 159, 262, 274
Hussein, Saddam, 68, 83, 88, 107, 121, 167, 173, 193, 211, 240, 245–47

Iceland, 35, 253
IFF (identification, friend or foe), 93, 96
Iguana 70/MH-3, USAF (I019), 109
I-Hawk missile batteries, 69
III Corps (Iraqi), 215
imagery intelligence, 106
INS (inertial navigation system), 24, 66, 113, 147, 149–50, 153, 225, 256–57
Incirlik Air Base (AB), Adana, Turkey, 74, 85, 102–3
Indian Ocean, 180
Initiative 17, 26
Inspector General's office, 23
Internal Look, war game/exercise, 44, 252
Iran, 18–20, 28, 44, 46, 63, 121, 155, 176–77, 274
Iran-Iraq War, 33, 173

Tractors on the Farm
Push and Pull

by Linda Ward Beech

Contents

Science Vocabulary 4

Tractors Move 8

Pushes and Pulls 10

Ways Objects Move 16

Find the Pushes and Pulls . . 24

Magnets Move Objects . . . 26

Conclusion 28

Share and Compare 29

Science Career 30

Index 32

Science Vocabulary

force
A **force** is a push or a pull.

This tractor uses **force** to move.

push

When you **push** something, you move it away from you.

This tractor **pushes** grass into a pile.

pull

When you **pull** something, you move it toward you.

This tractor **pulls** a cart full of pumpkins.

direction

A **direction** is the path an object takes.

This tractor turns to change **direction.**

motion
When an object is moving, it is in **motion.**

My Science Vocabulary

| direction |
| force |
| motion |
| pull |
| push |

This tractor is moving. It is in **motion.**

Tractors Move

This tractor is moving.

The farmer drives it across the field.

Pushes and Pulls

Tractors use **force** to move things.

force

A **force** is a push or a pull.

They can **push** and **pull.**
This tractor pulls a farm machine.

push
When you **push** something, you move it away from you.

pull
When you **pull** something, you move it toward you.

This tractor pulls pumpkins.

This tractor pushes grass.

push

Pushes and pulls can make things start to move.

They can also make things stop moving.

The tractor stops at a road.

Pushes and pulls can make tractors change **direction**.

The tractor turns onto a road.

direction

A **direction** is the path an object takes.

Ways Objects Move

This tractor is moving.

It is in **motion**.

The tractor moves fast on the smooth road.

motion

When an object is moving, it is in **motion**.

Tractors can move fast.
They can also move slowly.

The tractor moves slowly over a rough surface.

Objects can move in different ways.
They can move in a straight line.

This tractor moves down the field.
It cuts crops in a straight line.

The tractor's wheels push on the ground to turn.
Round-and-round they go.

Wind can push the windmill.
The windmill turns round-and-round, too.

The tire swing moves back-and-forth.

Tractors can zigzag.
They made this pattern.

zigzag

This farmland has a zigzag pattern.

Find the Pushes and Pulls

This picture shows a farm.

Name the pushes and pulls you see.

Two are labeled for you.